X

PRAISE FOR

TWENTYNINE PALMS

"Stark observations of beer and crack-fueled parties among jarheads and townies. . . . A fascinating, if chilling, read." —*Maxim*

"A book as stunning as it is shocking. In Stillman's prose, you can almost feel the blistering heat of the Mojave Desert, smell the stale beer of the seedy bars. . . . This book is unforgettable."
—*Arizona Daily Star Sunday Book Review*

"One of the best stories to come out of the high California desert. . . . A terrific book, and an important one."
—Judith Freeman, author of *A Desert of Pure Feeling*

"Stillman writes a dust-devil tale of murder, madness, and the military soaked in white-hot passion and razor-sharp insight."
—Lucian Truscott, author of *Dress Gray*

"This haunting desert-gothic tale [is] . . . an irresistible plunge into a palm-fringed nightmare."
—Ron Rosenbaum, author of *Explaining Hitler*

"Blows readers away . . . with the blood and guts and heart of the murder." —*Austin Chronicle*

"Good news for readers who've been waiting for a book to match *In Cold Blood*." —*East Bay Express*

"Stillman immerses herself and her reader in a world at which few desire even to peek, depriving both of cliché." —*LA Weekly*

"[Stillman's] writing and reporting skills are excellent."
—*Sunday Oregonian*

© 2001 by Martin Sugarman

About the Author

DEANNE STILLMAN is a former columnist for *Buzz* magazine. Her writing has also appeared in the *New York Times*, *Los Angeles Times*, *Rolling Stone*, *GQ*, *Village Voice*, and the on-line publications *Salon* and *Slate*. Her article in *Los Angeles* magazine, upon which this book is based, won the 1996 Maggie Award for Best News Story. Her work has been widely anthologized and her plays have won prizes in theater festivals across the country. Originally from Cleveland, Ohio, she now lives in Los Angeles, California.

TWENTYNINE PALMS

A True Story of Murder, Marines, and the Mojave

Deanne Stillman

Perennial
An Imprint of HarperCollins*Publishers*

Grateful acknowledgment is given for permission to reprint lyrics from:
"Do Me" by Bell Biv Devoe, © 1990 Universal-Unicity Music Inc., a division of Universal Studios, Inc. (ASCAP/BMI). International copyright secured. All rights reserved.
"Desolation Row" by Bob Dylan. Copyright © 1965 by Warner Bros. Music, Copyright Renewed 1993 by Special Rider Music. All rights reserved. International copyright secured. Reprinted by permission.
THE LADY FROM 29 PALMS, words and music by Allie Wrubel. Copyright © 1947 by Music Sales Corporation (ASCAP). International copyright secured. All rights reserved. Reprinted by permission.

A hardcover edition of this book was published in 2001 by William Morrow, an imprint of HarperCollins Publishers.

HarperCollins books may be purchased for educational, business, or sales promotional use. For information please write: Special Markets Department, HarperCollins Publishers Inc., 10 East 53rd Street, New York, NY 10022.

First Perennial edition published 2002.

Designed by Paula Russell Szafranski

The Library of Congress has catalogued the hardcover edition as follows:
Stillman, Deanne.
Twentynine Palms: a true story of murder, Marines, and the Mojave / Deanne Stillman.
p. cm.
ISBN 0-380-97560-2
1. Murder—California—Twentynine Palms—Case studies.
I. Title.
HV6534.T944 S75 2000
364.15'23'0979495—dc21 00-056002

ISBN 0-380-79401-2 (pbk.)

02 03 04 05 06 ❖/RRD 10 9 8 7 6 5 4 3 2 1

Contents

Introduction

On September 22, 1997, Debie McMaster got on a 7:30 A.M. flight at Chicago's O'Hare Airport. Thin, wiry, louder than Led Zeppelin when necessary, and when not, simply outspoken, on this day she was unusually subdued. During the three-hour flight to Ontario, California, she chewed twelve sticks of gum, drank three bourbons-and-Coke, read the horoscope for Leo—hers and her daughter Mandi's—in the current *Cosmopolitan* magazine until she had it memorized, and then was overcome with a serious migraine shortly before the plane landed. She had been feeling a slight sense of relief lately, and to those who knew her best, it showed. She had gotten her long red hair restyled, shaping it with bangs and a smooth, urban edge. The captivating flash in her famous green cat's eyes—dulled by years of exhaustion—had returned. Those who are schooled in the ways of selling say that to make an impression, the salesman, or really anyone for that matter, must hold a stranger in his or her gaze for ten seconds. Debie's first encounters generally lasted fifteen seconds, if one counted those things—such was the power of her regard, such was the kind of gift often given, it seemed, to those who had few other methods of persuasion. Her life force was flowing again because finally there would be a trial. Maybe it would be over by Christmas and she could start celebrating the winter holidays again. Maybe the world would stop blaming her for Mandi's death, maybe now they would see what really happened. But as soon as the plane began descending, her eyes blinked and stayed closed a beat too long. It was the first sign of one of her chronic headaches—sensitivity to light, light which poured through the plane's windows as it left the smoky brown haze of particulate smog behind, in this case, the strange, heightened light of a Southern California autumn, a light which on certain days is so clear it can slice through gristle and bone all the way to the heart of you, and things, and when the plane touched down, Debie kept her eyes shut and tried to ride out the pounding waves in her head. In a few days she would be sitting in court, a few feet away from the man accused of killing Mandi and her friend Rosie, watching him joke with bailiffs, fighting off eons of a blood legacy which whispered: *do anything but sit and watch*. But sit and watch she would do, she promised herself, pay attention to the gruesome crime-scene evidence—autopsy photographs, close-ups of wounds, the weapon itself—that would be presented in court, relive Mandi's last moments with her, summoning strength from the memory of Mandi and the need to see this through. The worst of it would be over in the first few days, she kept telling herself—the district attorney had promised her he would present the most horrifying photos as soon as he could, and would always signal to her before he unveiled them, or anything else of an upsetting nature, in case she wanted to leave the courtroom. He had also promised her that she would be one of the first witnesses called to testify—she had testimony that was relevant to the time line of the murders. Although she had faced the accused killer in court at a pretrial hearing in 1994 in which she was also called as a witness (and he had removed his glasses when she took the stand), this would be the first time that she would sit in the box before him for more than a few minutes, and after that, she would be sitting in the same courtroom until the trial was over, her eyes flashing a fire of grief and hatred and sadness without fathom, and she would watch and wait for the moment that she could deliver a *statement*—yes, at least she had that—a *statement* that in recent years had become a strange kind of cultural narrative, a do-it-yourself expression of what life had become without a beloved daughter or son or parent, of what it was like—and how much Debie hated saying this—as the victim of a crime. For weeks she had been keeping herself afloat with the promise of this and all the other little comforts, but as she walked across the tarmac at the Ontario Airport trailing her Tourister on wheels, she caught sight of a couple of fan palms and the tears started down her face. These were not tears like last time, when she knew she was going home at the first sight of Twentynine Palms, knew in her gut that she was back in the place that was ocean before desert, sorry, strangely hopeful cradle of all life everywhere—but tears of ache and failure and dread, tears that she was ashamed of having, for she knew she was back in the desert, which had reclaimed her daughter, and this time, it did not feel like home at all.

PROLOGUE:

Prelude to a Kill

The concern here is the Mojave Desert, the dry, baptismal font of national consciousness, mythological birthplace of America. It takes a big, white-hearted desert to fuel the pursuit of happiness, vast stretches of emptiness to suggest that the world can be possessed like an oyster, extreme tableaux of beauty to obliterate all memory of bad news. "Have a nice day!" the Mojave Desert tells the crossing parade—the Donner Party, the seekers of buried treasure, the cowboys, the ranchers, the people who rush for Hollywood gold—"Good luck! Think positive!"

Called the Mojave Desert after the Indians who once lived there, this blank, sunny slate bears a name that has defied the plundering of linguists, the meaning of the original term, *hamakhaav,* long ago swept away by the Santa Ana winds, that strange atmospheric condition born in the desert which raises the skin on all living creatures and is said to warn of earthquakes. But the mysterious name fits; the unknowable is unnamable, too. The Mojave was here before California, Nevada, and Arizona planted their flags in it, and it will be here tomorrow. Not that it's keeping track of time—history doesn't matter out here; it's space that counts, space that drives the country, space that suggests the possibility of declaring bankruptcy and starting over somewhere else, space that maintains the illusion of hitting the jackpot on some get-rich-quick scheme, space that whispers, *make bombs and bring down the government all by yourself.* In a weird bakery of the impossible, a vast scape of tortured beauty where all things are equal and do what is necessary to survive, personal demons aren't demons at all, but just some other creatures who need a drink.

Senseless violence, the world calls it, but the Mojave knows otherwise. The Mojave knows, has always known, that the violence is not senseless, the disturbing acts that unfold on its sandy stage in fact make perfect sense. For that is the very nature of the place, to convey meaning, to show events in living color on a giant screen in bas-relief, to make it seem as if everything is happening for the first time, even if for some, it is the last, or simply the latest in an endless spiral of repetitive, nowhere acts. And that is the nature of the people who come here. They are starting over in the oven of American Zen, refracting into new souls with each infinitesimal turn of the earth, cranking up the Van Halen as the sun becomes the moon, being right here, right now, this is it, but Officer, last chance for new ID.

But the concern is not really with all of the Mojave, just a part of it—one aspect of its character—its very heart. This is the town called Twentynine Palms, which is found at an elevation of four thousand feet at a longitude of 34°08'09" North and latitude of 116°03'15" West, one hundred and eighty miles east of Los Angeles, a short distance but a very long way. Its stage props are the tortured rocks and freak-show plants of its progenitor, but it is a heightened version of the Mojave; from it the Mojave might have been cloned. It sits on top of seven known fault lines and perhaps countless undetected cracks in the earth. The bottomless fissures crisscross and zigzag for hundreds of miles in every direction, creating the most volatile web of geography in the American West, a region geologists call the Eastern

California Shear Zone. To the north and east runs the Emerson Fault, epicenter of the 4.5 Emerson Quake in 1975. To the west run the Galway Lake Fault Zone and the Pinto Mountain Fault, site of nonstop temblors ranging from one-pointers, which are imperceptible to all but the most highly attuned desert creatures, to jolting slip-and-fall three- and four-pointers, which make for a noisy response among cactus wrens and mourning doves and send jackrabbits skittering across the sands and collapse the fragile nests of the desert tortoise and snap its freshly laid eggs in two. To the south and east run the Cleghorn Lake Fault and Homestead Valley Fault; these two cracks in the earth met fiercely tens of thousands of years ago, and they have continued to collide with each other so violently and so frequently that they have shaken and thrust upward the Coxscomb Mountains—a peculiar range outside of town that always looks bruised. In 1992 the intersection of the Cleghorn Lake and Homestead Valley faults ruptured in a quake of 7.2 magnitude, epicentered near Twentynine Palms in the town of Landers. As the ground in the Eastern California Shear Zone fell away, the Coxscombs lurched skyward—some say the ancient peaks gained two inches in the blink of a raven's eye. The Landers Quake echoed across the West, at the beach in Santa Monica where the palm trees swayed in response to the distant ground shivers, in Las Vegas where the casinos blacked out for a moment, hinting that there might be such a thing as time, in Montana where a truck drove off a two-lane, and in New Mexico where nervous desert dwellers in white helmets checked and double-checked missile silos that seemed relics of a distant global configuration. In Twentynine Palms, some residents were so alarmed by the force of the quake that they did not sleep inside—under a roof—for days. Was the Mojave Desert beginning to eject its latest squatters, reclaim itself? Perhaps so—in one way or another, every so often, perhaps when it tires of its own stillness, it likes to scare people away, to writhe in pain and shake uncontrollably in delight, to stir things up, to make people think—otherwise how can its treasures be calipered, appreciated?

And then there are the times when the Mojave Desert gets serious, wants more than fear and awe, demands a blood sacrifice. The personal-rights party has gone too far. Things must happen. Often, a girl is involved. Often, some boys. Generally, a knife. And then there is the military. In this case, the few, the proud . . . the Marines. The blood must flow, attention must be paid: the desert says, "Don't tread on me, I'm where the party started and one of these days, I might just shut the whole thing down."

PART ONE

In the Beginning

And you, Lord, were glad to have her there: you were glad there were women in the world, with their hair and their tears: women who know how to weep better than men; and this is a richness they have in their eyes, a sign of humble and sweet remission such as men rarely know.

Luke 7:44–7, modern interpretation

The miner came in forty-nine,
The whores in fifty-one;
And when they got together,
They made the Native Son.

anonymous rhyme

Six years had passed. The double homicide case of *The People of the State of California* v. *Valentine Underwood* went through five judges, dozens of delays, three changes of venue. It became the longest-running untried criminal case in the history of San Bernardino County. Everything that could possibly visit a murder case visited this one. A judge retired. The San Bernardino County Court system ran out of money and could not schedule its many murder trials. The case was moved from Rancho Cucamonga, near the West Valley Correctional Center where the alleged killer was housed, to the high desert town of Barstow (a joke word in old Johnny Carson monologues and the town where infamous killers Perry Smith and Dick Hickok laughed at their own grim private joke when a motorist passed them by as they stood on the freeway, noting the probability that they would have murdered him if he had stopped to pick them up), and then finally to Victorville, a stop on the old Route 66, near Barstow and bigger, more able to handle murder cases that generated a lot of paperwork. The defendant's first lawyer, court-appointed Mark Sullivan, once called the Perry Mason of Palm Springs by *Palm Springs* magazine, was replaced. The defendant's second lawyer, pricey, small, scrappy hired gun Garrett Zelen out of L.A., received an immediate six-month postponement so he could get up to speed on the case. When he did, he mounted a vigorous and flashy defense, arriving at the Barstow courthouse where the case was originally to be tried in a black Porsche and Armani suit, not endearing himself to locals in this land of Fords and Chevys and all other things American, outstyling Gary Bailey, the modestly dressed district attorney, also scrappy and small. Zelen launched what turned into a nearly three-year campaign, filing dozens of motions, arguing under citations of cases with names like *Schlecknopf-Bustamante* and the evocatively named Old Testament throwback, "Fruit of the Poison Tree statute." His quiver included a motion contesting the arrest of his client based on his contention that the arresting officers, employees of the San Bernardino County Sheriff's Department, did not have authority to make the arrest, considering that the location of the arrest, the Marine base at Balboa in San Diego, was a federal reservation; a motion that argued for the dismissal of the case based on the fact that the defendant's Miranda Rights were violated—fifty-one times (there was indeed a violation but the case was not dismissed; rather, statements in which the defendant placed himself at the scene of the crime after the murders were not entered as evidence at trial); a motion that contended evidence such as the defendant's blood was seized illegally since he wasn't under official arrest when it was drawn; a motion that argued that the blood was tainted due to the possibility that one of the arresting policemen kept the vials in his refrigerator next to a six-pack of beer; a subpoena for a writer's notes, based on the contention that the defendant's right to a fair trial under the Sixth Amendment trumped the reporter's freedom-of-the-press rights under the First Amendment because the writer might have come across information which exonerated the defendant; a motion which resulted in the calling of the ultimate presiding judge's wife—coincidentally the court clerk—to the witness stand, to answer the assertion that the judge had spoken to her about the case and she in turn had discussed it at a party in Victorville; a motion contesting the accuracy of DNA evidence which allegedly linked the accused client to the crime scene (at the time of the filing of this case, this was only the second case in San Bernardino County to use the then-controversial process of identification, and one of the first in the country); a motion alleging that the arresting cops were racist, and part of an overall policy in San Bernardino County that targeted black people; a motion suggesting that jury selection in San Bernardino County was racist, culling a disproportionate number of white people; motions for various delays due to scheduling conflicts with other cases, one of which involved the defense lawyer's own trial for obstruction of justice in Los Angeles (he was accused of interfering with a police sting operation against a client, but the case was dismissed); and finally, in 1996, after he had lost virtually every single motion except for requests for delays and the case was about to go to trial, a motion to have himself dismissed. He had apparently not informed his client of his impending departure and upon hearing the news in court, the long-untried and incarcerated defendant, in ankle chains and high-security, orange prison jumpsuit, a foot taller than Zelen and possibly twice as wide, leaped to his feet and lunged for his attorney. Two bailiffs dragged him off. The judge granted the motion; for the first time since he had represented this particular client, Garrett Zelen had won a major argument. And so the case came full circle: a new attorney was appointed by the court. His name was John Hardy, a stocky World War II veteran with a leathery face whose left arm was an old-fashioned prosthetic device in the form of a hook. He had practiced law in the Mojave for years, defending many a desert killer. For the next few months, he would be busy with another case involving the military-related murder of a Twenty-nine Palms local. Coincidentally, a witness in that case would also be called to testify against Underwood. To get up to speed on his latest case, and the thousands of pages of documents already on the record, Hardy asked for, and was granted, a one-year postponement.

At the baggage-claim area, as had been arranged, Debie was met by Jesse R. Fulbright, the victim witness advocate who worked in the San Bernardino County Victim Witness Assistance Program. A former Marine who had served in Vietnam, Jesse, physically speaking, would no longer pass muster. He was five feet ten inches tall and weighed 360 pounds. But his size seemed to suit this job of comforting the victims of terrible crimes who were witnesses at the trial of the person accused of killing their loved ones; a person could get lost inside Jesse's huge hug, almost go swimming in his big, sad eyes. Had he become big before being hired as a victim witness advocate? Or had he—as American presidents do over time—absorbed and physically manifested the marks of the job, in his case the deep wells of sorrow which those he counseled poured into his very flesh? At first sight of Debie, he stretched out his arms and wrapped her up and told her everything would be okay—difficult, but okay. Then he picked up her other suitcase, the larger one she had bought on special at Kmart just for this trip, and they drove into the high desert toward Victorville.

Jesse helped Debie get checked in at the Best Western. The desk clerks knew him, and they knew the routine of checking in witnesses, knew how distraught or nervous some of them were, and they signed Debie in quickly and in soft tones, like admissions clerks in a hospital before major surgery. They refrained from saying, "Hope you enjoy your stay at the Best Western," offering simply their assistance should Debie need anything, night or day. Debie told them that as a matter of fact, there was something they could do. Jesse had mentioned on the way to Victorville that he was returning to another airport to pick up two other witnesses—Timothy Carmichael, who had found Rosie and Mandi's bodies, and Trenton Draper, who had spent time with Rosie and Mandi before they were killed. "Could you please give Carmichael this note?" Debie said, scribbling down a request for him to call her as soon as he got in. A clerk said that she surely would pass the note on. Jesse escorted Debie to her room, admonishing her not to upset herself with unnecessary meetings, and Debie said that she felt she should spend some time with the people who last saw Mandi, she didn't know why, it just seemed like the right thing to do. Her room was a double in the smoking section on the second floor, facing the interstate, and immediately below, a parking lot with a couple of big rigs, an RV, a few Harleys, and assorted cars, mostly American. Jesse apologized for the view, explaining that it was the best he could do on the state's budget, then said he would be back the following morning at eight-thirty to pick Debie up and take her to court. "Thanks, Jesse," Debie said. "You're an angel." He said that he was just doing his job and then he shambled down the cement stairway to the parking lot and drove off in his aging beige Honda sedan. Debie unpacked, smoothed out the wrinkles of the outfit she planned to wear to court on the first day of the trial, hung it on its special white satin hanger, took a couple of extra-strength Tylenols, and showered. She called Chicago, to tell her twenty-one-year-old son Jason and her longtime boyfriend Mike Ramirez that she was okay. As the sun sank into the Mojave, she closed the curtains, lay down, breathed into the never-ending stream of freeway suck-and-whine, and went to sleep.

Sonora was the most violent of all of the California gold-rush mining camps. Located between Whiskey Flat, Angels Camp, and Columbia to the north, and Jamestown and Mariposa to the south, the town originally consisted of a few tents and shacks, like the neighboring communities along the spine of the Sierra Nevada mountains. Those who came to live here had started the journey with high hopes and a voracious appetite, following the setting sun with provisions and possessions that were dear, and picks, chisels, hammers, axes, shovels, and anything else that would help them take treasure from the land, including massive gold extraction machines (useless inventions that were hawked to the gold-fevered). By the time they arrived in Sonora, many fortune hunters had endured hardship beyond conception. Some had come from afar on decrepit vessels that had sailed from eastern ports for South America and then around the treacherous Cape Horn, and northward through the Pacific for San Francisco, where, weary from malaria, dysentery, and scurvy, they stumbled off and headed inland.

"Men paced the decks in desperation," one argonaut wrote, "climbed the rigging, or lay prostate in their bunks staring at the bulkhead. They began to gamble and to drink. Some went insane and had to be shackled in a storeroom for the remainder of the voyage, their California adventure over before it started."

Others traversed harsh overland trails like the Santa Fe, the Texas, the Mexican, or the Hastings Cut, a shortcut off the Old Emigrant Trail leading right over the mountains into El Dorado. In 1848, the Donner Party perished on this time-saver, trapped for the winter by an early blizzard. There were some among this wagon train who did not turn left at the Cut, who thought that the little-known route was too risky. After an argument which nearly came to blows, they split off from the Donner family and continued on the longer trail for Oregon. This group, those who turned right at Hastings Cut, included Alexander McMaster. His clan emigrated from Ireland to Maryland in 1750, and within a generation began making its way to the opposite side of the country. The family's first Western outpost was Springfield, Illinois, where Alexander worked as a bodyguard for Abraham Lincoln, beginning a family tradition of service to country. When Lincoln embarked on his presidential campaign, McMaster decided to seek his fortune west of the Mississippi. After several years in Oregon, he settled in Sonora. Debie McMaster—his great-great-granddaughter—always thought of him as the one guy in the family who did something smart.

At the time when Alexander McMaster arrived in Sonora, there had been a series of hangings, knifings, and shootings. The code was Miner's Law, and by it, everybody lived and died; by it, they raised their kids. McMaster married the daughter of another gold digger and they had two sons. The family found enough gold to move to a small shack. By the late 1890s, the gold was gone, and those who remained turned to other endeavors. James, the eldest son of Alexander, married into a family of local growers, the Dentones, who had come one by one from Genoa on a series of treacherous oceanic crossings in the 1870s. By the early 1900s, after finding some gold on their land and doing well with their small tomato farm, they had become successful—wealthy by some standards—and regarded themselves as one of the founding families of Sonora. A son, Clarence McMaster, married his high-school sweetheart, Rose Inks. Rose's family, too, were among the founding families of the town—her great-great-grandfather, a prospector from Sicily, had married a Miwok Indian long before Alexander McMaster had crossed the Rockies. But the Inks family does not like to boast about their presence among the early settlers in Sonora. In fact, they don't like to talk about it at all. Rose's family, especially the women, seemed to have been hounded across time and space by a jackal from which there was no escape.

"I don't know much about my ancestors," said Rose Inks—now Powell—as she pulled a Newport from a pack and lit it up. It was four o'clock in the afternoon at Dave's Coffee Shop in Oakland. Soon she would be starting her second pack of the day. "I wish I knew more but I don't," she continued. "I remember hearing that my grandfather built the first silent movie house in Denver. I never knew my father's parents. Excuse me." The seventy-two-year-old waitress moved down the counter to pour a coffee refill. "I wish I knew more but I don't," she said. A couple of patrons nodded. "Anyone else?" she called. An old guy with an old-guy stubble alone in a window booth indicated yes. "You got it," Rose said. "I'm not particularly close with my family," she said from across the coffee shop, topping off the cup. "I wish we would see each other more often, but we don't." She returned to the visitor, picked up her cigarette from the ashtray. "Oh well," she said on the inhale, and then blowing it out, "that's just the way things panned out." Just like so many of the defunct mines in mother-lode country, Rose's home turf. Rose Powell, once a beauty with black hair and black eyes, still looks good, weary but good, has the kind of muscular toning that comes not from gyms but physical labor, looks the way people who were born to hard work look when they are old—fit, signed off, tired. "They'll have to carry me outta here with a toe tag," Rose said. "I like Dave's."

Rose has been working at Dave's for the past twenty-seven years. In the beginning, it was full shifts, every day, 6:00 A.M. to 2:00 P.M. Now she is down to weekends, driving the one-hour round-trip from home every Saturday and Sunday in her 1979

Camaro; the income supplements her social security and occasional cash from baby-sitting for the children of a fireman in her apartment complex. "It's my old stomping grounds," Rose said. The coffee shop is on the corner of Forty-second and Broadway in Oakland, in an ancient working-class neighborhood, across the street from Oakland Stadium ("I once met Vida Blue," she said, and then said he gave her a photo that he inscribed, "To Rose, Good Luck on the Road of Life, Vida Blue"). "I would rather be here than at home," Rose said. She lives in a low-income housing development in the town of Pleasant Hill with a cat named Tom. Her one-room residence is a few hours from Sonora, where she was born. She misses Sonora: her mother, Lorena, is buried there, and they were very close, she recalled, although the closeness was not derived from a series of storybook moments, memorable celebrations, or happy milestones. "The drive is too much for me now," Rose said. "I have high blood pressure. I get these palpitations. I don't want to have a heart attack on the road." At five o'clock, the night shift arrived. Rose cleaned up her station and gathered her things. " 'Night, Rose," the old guy at the window booth called out. "Oh, I'll be back," Rose said. "You know me—I love to work. If Dave would let me, I'd work a double shift." It is an ethic, and an oft-repeated refrain, that has come right out of the boozing, brawling, beautiful Sierra Nevadas. Rose joined the visitor in a booth. "You probably heard by now that my parents were the town alkies?" she said, and quickly looked away. "But my father—he never missed a day of work."

On October 3, 1928, Lorena Inks was about to give birth. She was alone at two in the morning in a two-room shack in Sonora. There was no telephone. Her contractions were two minutes apart. She was scared—the baby wasn't due for another two weeks. Her husband James, a bulldozerman for a government agency, was away on a job. The sky was ablaze with all the constellations; the coyotes howled. As the contractions came faster, Lorena ran for a towel, grabbed a knife, then squatted on the living-room floor and began to pump out her first child. The infant was slow to emerge, and Lorena screamed louder than all of the night creatures outside the tiny mountain shack, and all the night creatures screamed back and to each other, for a long time it seemed, until it was over. She cut the umbilical cord and hugged her new baby girl. Girls in California could do anything they wanted. There were no rules; a girl could stake her own claim in whatever form that took and there was no one to stop her. Lorena held her baby, then lay still for a moment, exhausted. James kicked open the screen door and walked in. He reeked of whiskey, his entire being oozed liquor: oh yes, people said he was the finest bulldozerman in the county, but Lorena knew that he was a bulldozerman who wished that things were as they were not, wished that he had more than a fifth-grade education, wished that he had anything resembling a choice in the matter of his life. "What the hell's going on?" he shouted. "What are you doing down there?" Lorena showed him their child. "Meet Rose Marie," she said. It was the name of her great-grandmother on her mother's side, or maybe it was her father's side, she couldn't quite remember, it was the one who had come to these shores from Sicily—or was it Spain?—in the 1860s. James stormed out, and did not return for three days.

They called her Rose for short. Like her mother, Rose was spunky and independent. She liked to visit her grandmother, who had a radio in her house. Sometimes they would lose themselves in broadcasts of *The Lone Ranger*. At home, Lorena always had a glass of something in her hand. Rose didn't know exactly what it was, she only knew that it made her mother happy, she would stop crying when James left, which he did more and more frequently now that there were two additional children in the house. Whenever he would come home, if he wasn't drunk already, he would get drunk and so would Lorena. They would fight and he would beat his wife. "Stop it, Dad!" Rose would cry. "Leave Mom alone!" When the fight was over, James would leave again. "Here, Mom," Rose said, over and over again. "Have a drink." One day, when Rose was twelve, Lorena ran away from her husband and her three children, from Sonora, the mountains, down to the sea, on to San Francisco. Every day she sent Rose a letter. Every day Rose would stand at the mailbox in front of her house and wait. As soon as the mailman handed it to her, she would tear it open and read, pacing up and down the dirt driveway. "Dear Rose," Lorena would say. "I miss you very much. Don't worry about me, I am doing fine. Please look after your brother and sister and try not to get into fights with your dad. I love you—Mother."

One morning a couple of months later, Rose Marie put on her best plaid skirt, her white blouse with the Peter Pan collar, and her scuffy saddle shoes with the pink laces. She cracked open her piggy bank and counted out all the quarters and nickels, dimes and pennies. She removed the stack of letters from her secret place under her bed, stuffed them into her book bag, headed out, and walked right past school, directly for the bus station. "One ticket for San Francisco," she told the cashier. "Does your mother know about this?" asked the man in the booth. "Oh yes," Rose replied. "She's waiting for me." The little girl boarded the next bus and, three hours later, emerged in San Francisco and hailed a cab. "Where to, girlie?" the cabbie said. Rose retrieved the letters from her school bag, so neatly tied with red ribbon, and showed the cabbie the address. "You sure you wanna go there?" he said. "Oh yes," Rose replied. "That's where my mother lives." A little while later, the cab turned down a small alley in the red-light district and stopped in front of a flophouse. "Okay, girlie, this is it," he said. Rose, who was used to squalor, although not like this, tried to look unfazed. She reached into her purse and took out her plastic wallet. When she opened it to retrieve her change, the cabbie spotted a cracked black-and-white snapshot of toddler Rose and her mother standing under a giant sequoia. "On the house," the cabbie said. Rose got out, stepping quickly over a couple of winos who were splayed across the entrance.

"Be careful," the cabbie called out. But Rose had already disappeared into the shabby hotel, and was running up the stairs to room number eight. She knocked anxiously on the door. "Who is it?" came her mother's voice. "Mom!" she called. "It's me!" Lorena stumbled to the door and Rose heard her fumble with the lock. "Mom, are you okay?" she said. Finally Lorena was able to unbolt the door. Rose ran into her arms, too fast to notice how disheveled her mother was, too happy to notice the empty bottles of wine all over the room, too lost in comfort to notice that her mother had been drinking for days. "Mom, can I stay with you?" Rose said. Lorena burst into tears. She knew that she was not a fit parent. "Maybe someday," she said, and hugged her daughter. Then she called James. James called the police. Within two hours, the juvenile authorities came to fetch Rose. She would not let go of her mother. "Rose, everything will be fine, I promise." Rose kicked and screamed and flailed. Lorena held her until she was spent and then the two men took her away. In the car as they drove back to the mountains Rose asked them to stop. She had forgotten the letters and wanted to get them. The men told her they would be safe with her mother. Rose did not say anything else for the rest of the trip, all the way back into the higher elevations, and she did not say anything for the entire week that she was placed in a detention home. After that, she was sent to live with her aunt in Angels Camp—another two-room affair, but better kept, and not so raucous:

Aunt Mildred lived alone and did not drink. For a long time Rose Marie lost her appetite. She was skinny, ten pounds below average for a twelve-year-old—and it showed, because at five-seven, she was tall for her age. She was not doing well in school. Often she would put her head down on her desk and sleep through the school day. One morning she woke up with a temperature of 102. The doctor was called. He could find nothing wrong, other than a nervous condition. It was not the first time that this diagnosis had been made, but it was the first time Rose's temperature had climbed this high for no apparent reason. Aunt Mildred tried to soothe her trembling niece with a cool washcloth across her forehead. It was time to pass on an important family story—the Time Jesse James Stopped at Great-Grandma Eileen's House. "One day, Jesse and his brother Frank galloped up to the old homestead outside of Abilene," Aunt Mildred said. "Great-Grandma couldn't believe her eyes. She was mighty scared. She told Great-Grandpa Pete to get his rifle. But before he could, Jesse said, 'Don't worry, Pops, we just want to pass the evening with y'all, partake of some hospitality, then we'll be on our way.' So Great-Grandma and Great-Grandpa invited the James brothers into their house, which they had erected themselves with their bare hands, and Jesse and Frank admired their handiwork. 'Ma'am,' Frank said, 'we're mighty tired.' So Great-Grandma Eileen showed them a spare bedroom. It was already made up because that was the way they were, and Jesse and Frank went in and hit the hay, just like that. In the morning, at the crack of dawn, the James brothers got up and came to the table for breakfast. Great-Grandma Eileen had laid out a fine breakfast, as she always did, company or no. She made her trademark hotcakes and sausages and her coffee, which was known all over the prairie and some say the plains, too. 'Mighty tasty,' Jesse said as he and his brother cleaned their plates. 'You're not going to kill us, are you?' Great-Grandpa

Pete said. 'Pops, don't you worry,' the boys told him. 'We just needed to rejuvenate ourselves, and now we'll be on our way.' Jesse and Frank got up, tossed a shiny gold coin minted by the U.S. government onto the table, tipped their hats, bade a fond farewell, and galloped off over the horizon." Then Aunt Mildred reached into her apron pocket and flashed something round and glittery through the air. "Did you ever see such a shiny coin?" she said. A few days later, Rose's fever subsided and she went back to school, bought off temporarily by this tale of her family's encounter with history, but knowing somewhere in her blood, her cells, that it was nothing more than that really, a tale told by those who count for nothing.

"Will you marry me?" The words gave Rose Inks a thrill down to her toes. She was sixteen and an usher at the Alhambra Theatre in Sonora. As she showed her handsome and cocky suitor to his seat, she laughed off the proposal. "I'm serious," said Clarence McMaster. "Serious as a heart attack." "I don't know," she said. The lights went down and the movie began; it was *Johnny Guitar*. Yes, I'll marry you, she thought as Joan Crawford collapsed into the arms of Sterling Hayden. But I won't say yes until you ask me again ten times. It was just a little game she wanted to play. She wanted to find out if Clarence really was as serious as a heart attack. A few days later, Clarence returned to the Alhambra. He bought his ticket for the matinee and waited for his favorite usher to show him to his seat. "Back for *Johnny Guitar*?" Rose said. "I'm a sucker for westerns," Clarence replied. Rose replied that she was, too. Then Clarence asked, "Rose Marie Inks, will you marry me?" Rose forgot the promise she had made to herself and said yes without a moment's hesitation. Clarence was such a handsome seventeen-year-old—why play hard to get when she was just going to give in anyway? Maybe they could start living together right away. She hoped they could. Were they even allowed to get married? She hoped so, because she was once again living with her father, and when he got very drunk, he beat her.

A few days later, Clarence came to fetch Rose for a date. Rose fidgeted with herself before a small mirror next to the front door. Her father sat in the kitchen with a game of solitaire and a half-empty bottle of cheap whiskey. There was a knock. Rose quickly opened the door. Clarence held out a bouquet of summer lupine he had grabbed from the yard next door. But James looked the other way. Clarence followed Rose into the kitchen, where she put the flowers in a vase. "Clarence, these are so pretty. Dad, this is Clarence McMaster." James turned over a card and stared at Rose's suitor. "You can't marry my daughter," he said. "Dad!" Rose said. "Am I talking to you?" her father replied. Clarence promised that they would wait until they had finished high school. James poured himself another glass of whiskey and told Clarence to leave. Clarence replied that he and Rose had a date. James started to get up from his seat, then stumbled, but told them to get out. Rose grabbed a picnic basket she had prepared and left the house. Clarence offered his hand to James. But James held proudly on to the chair and told Clarence that if he knocked Rose up, he'd blow his head off.

In the car, Rose thought that she was pretty lucky: Clarence might have arrived after James's eighth belt of whiskey that afternoon, instead of the sixth or seventh, and then James would have been in a complete rage. She pulled out her pack of Newports, removed two, lit both, and handed one to Clarence. Clarence sucked in, looked at his reflection in the rearview mirror as he often did while in the car, smoothed back his thick black pompadour, and thought, Okay, one down, one to go—there was the matter of his family. He could handle them, he always did. There wasn't anybody he couldn't handle—just look at how that old slob James Inks couldn't even get up off the floor to say good-bye. He should thank his lucky stars that I'm taking his daughter the hell out of there. As the pair headed through the main intersection in downtown Sonora in Clarence's new Chevrolet coupe, past the general store that Clarence's grandmother owned, he turned to Rose and suggested that she come over right now and meet the McMasters. That way, they would never have to worry about it again. Rose agreed. She had warned Clarence about her father, and in turn, Clarence had warned Rose about his mother and aunt. Not that Clarence needed to be warned about James Inks. Like everybody else in town, the McMasters had nothing kind to say about the Inkses, especially Lorena, who was a local embarrassment. All of this to-ing and fro-ing between the families regarding their respective standings in the community only fueled the romance between Rose and Clarence; they were madly in love and the madness reached even greater heights whenever they thought about how much they pissed off their parents. But as Clarence went to unlock his front door Rose tried to stop him. She felt vaguely queasy. Clarence assured her that his family would not hate her. Rose was convinced they would say something terrible about her mother. Clarence opened the door and urged Rose into the living room, promising that the meeting would go smoothly. "You look just like Lorena," Clarence's mother said. "Has she quit drinking?"

In a grove of redwoods outside of town, they picnicked on peanut-butter-and-jelly sandwiches that Rose had made. During a cloudburst, they hid inside a giant knot in a redwood trunk. It was a warm summer afternoon, a moment of grace in lives inscribed for many moments of heartache. For the next two years they were inseparable. After high-school graduation, they married. Rose was two months pregnant. They moved into the two-room apartment above Clarence's mother. His grandparents lived next door. His cousins lived across the street. His sister and her husband and their kids lived around the corner. This would be her new family. They often stopped by to see how she was doing, always making sure to ask if Lorena had sobered up. Like you're all so high-and-mighty, Rose thought to herself. Like you're all so picture-perfect. I know what goes on in your house, I know how you treat your kids. So don't you try and get all snooty with me.

There was no work in Sonora, so Clarence found employment down the mountains at a 7UP bottling plant. He was gone during the births of his first two children, Marshall and Darrell. He was gone most of the time. On the evening of July 28, 1954, Rose Marie—nine months pregnant with her third child—felt her water break. "Goddammit!" she said as the fluid cascaded down her legs. "I knew he wouldn't be here!" She rushed out to her old Buick and drove herself to the High Sierras Emergency Clinic twenty miles away. The baby waited for her father before making her entrance two hours later, just as Clarence tore into the hospital driveway and raced down the hallway to the delivery room. Rose showed him their first daughter—Debra—and he admired his beautiful little girl. Then Rose turned aside with her child, knowing that Clarence had shown up because he happened to be driving home tonight and spotted her Buick in the clinic parking lot. They had not talked to each other in weeks.

Pregnant with her fourth child, Rose stood before an open casket at St. Mary's Cathedral in Sonora. She was twenty-six. She had not seen her mother since that day in San Francisco fourteen years ago. At forty-one, Lorena was dead, beaten to death in her hotel room. She lay there for a day, perhaps alive for a few hours after the beating, perhaps not, until a foul odor alerted a man in the room next door. When police arrived, they found the stack of letters from Lorena to Rose, still tied with red satin ribbon. The body was taken to the morgue. No investigation was mounted, no arrests were made, the police were looking for no one—the coroner said Lorena Inks had died of cirrhosis, but there was no official autopsy report and no police record. When Rose tried to get the letters back, she was told that they had disappeared. It was as if Lorena had never lived. Rose, now a grown woman who had stopped crying long ago, had learned that tears only led to more beatings. She wanted to cry as she stood with her mother, but could not. She was transfixed by all the bruises. Lorena's eyes were black-and-blue. Her arms had purple and yellow spots. She was so thin, she resembled a famine victim. Rose kissed her mother on the forehead and left Sonora for the last time. She remembered a letter Lorena had sent her long ago: "Don't worry about me, I'm doing fine." Yes, you are, Rose thought as she sucked on a Newport heading around a tight curve on Route 95 out of the mountains. If there was one thing that life had taught her, her mother was now doing just fine.

"Rose, why don't you and the kids come and live with me?" Clarence said. Two years had gone by. Linda, the fourth child, had been born. He was calling from the town of Alameda, near San Francisco. He had rented an apartment in a low-income housing project, hoping to lure Rose out of her hometown long before her mother had died. "We'll finally have our own bedroom," he said, "and there's one for the boys and one for the girls." It sounded good to Rose, even though she had the feeling that he had been having affairs, and disagreed with his brutal style of disciplining the children, which included severe spankings with a thick leather strap that came to be known as "the spanking belt." She wanted to get away from their small flat, and from Clarence's family, who owned it. When Rose and the children arrived a week later, Clarence was standing in front of their apartment to welcome them. His skin was yellow. "Didn't I tell you I was sick?" he said. Rose told him that he knew he didn't, and she wondered why his girlfriends had suddenly become unavailable. Clarence turned on the charm—"I didn't tell you because I figured why would you come after all this time just to do a bad boy like me a favor?" "I can't say what I would have done one way or the other," Rose said, trying to keep herself from becoming totally sucked back in, but knowing in her gut that sooner or later things between them would get ugly, just like they always did when Clarence reappeared, and bracing for the day—and it would come—when what she referred to as Clarence's "swave and dee-boner" act would melt into the next one down—crude and rude. But there was nothing she could do. Clarence had made the decision for her.

Clarence was hospitalized with yellow jaundice. Rose visited him every day with the kids. She tried to see past the tinted skin, to remember the striking seventeen year old she had fallen in love with in high school, but began to feel as if the illness carried a message. The high-school romance had turned to poison. A dutiful wife by day, Rose now danced the nights away, embarking on the girlhood that she never had, hooking up with big burly guys who brought toys for her kids and carried her off on motorcycles with loud pipes when the sun went down, and returning late—kissing them all good night but reminding them that she was married. Sometimes she would turn on the radio and play the country-and-western station, light up a Newport, take a smooth and satisfying drag, and sing along, alone in the dark, with Patsy Cline or Loretta Lynn. When Clarence checked out of the hospital, he confronted her, telling her, in front of Marshall, Darrell, Debie, and Linda that she was a tramp, just like her mother. He filed for divorce and was awarded custody of the children, arguing in court that Rose's choice in friends made her an unfit mother. Rose was unable to afford a good lawyer and successfully counter the argument with such information that Clarence was more than a "tough disciplinarian," as it was called at the time. So she was not surprised that suddenly she had lost her marriage and her kids—that's just the way life was for the women in her family.

"Your Honor, since I am of legal age to make such a request, I would like to live with my mother in Oakland," Debie McMaster said to the judge. She was sixteen and had been living with her father since the divorce, along with the other kids. She did not want to live with her father anymore and, in fact, had run away to her mother's. Of his two daughters, Debie explained, Clarence preferred Linda. Debie was the more complicated of the two, and the smartest of all four children, in every way as wily and cagey as her father, but she also had her mother's wild streak. Sometimes, when even the beatings with a leather belt couldn't control Debie, Clarence would send her across the street to live with an aunt. Sometimes he would demand that she come home. This time she was not going to: a week before her appearance in court, she had taken a particularly severe punishment and run away to her mother's. Her mother had immediately petitioned for custody. Debie explained to the judge that the incident happened after she had returned from school and displeased her stepmother, Jean. Jean hit Debie and Debie hit her back. Later, when Clarence returned, Jean told him that Debie had struck her. Clarence beat Debie with his fists. The next day Debie took the bus down the mountain to Oakland. "And so, Your Honor," she said, "that's why I want to live with my mother."

Like a migrating bird that follows an encoded path to a

stopover on the way home, Debie walked in the footsteps of Lorena and Rose, out of the mountains and down to the lower elevations, heading for a safe house and returning to the shelter of her mother's wings. When she arrived, she did not notice that her mother's wings were broken. Rose was now married for the third time, to a huge bear of a guy named Herb Powell. She had met him years ago in Sonora, had a crush on him, and having heard that he was in Oakland, looked him up when her short-lived second marriage to a drunk named Big Leroy ended. She was now working as a waitress and living in an old Victorian house with her son from her second marriage, Little Leroy. Rose wanted her children to have a father. When she and Herb got together at a local pub, Rose told him that she had a house. "Well, I'm a roofer," Herb said. "When do I start?" He winked and reached across the booth for her hand. Two days later, Herb came over and started working on Rose's roof. A day after that he moved in. Within a week they were married at city hall. Herb began hitting Rose shortly after the wedding, always making sure to keep his fists away from her face so people would not see the bruises. When Debie arrived, Herb and Rose had been married for eighteen months. They had a baby girl named Brenda. A house, a porch, a yard, a new neighborhood—Debie smiled when she saw it all for the first time from the backseat of the cab that she had taken from the bus station. She paid the driver with the last of her allowance money and hopped out, still in her school uniform, her "good outfit," and carrying her little powder-blue Tourister suitcase with her initials on top that her mother had scraped up the money for and sent her for the move. Rose ran down the steps from the front porch. Debie threw herself into her mother's arms, and Rose hugged her back, hard, happy, but in much pain, sharp physical pain from the two cracked ribs she had sustained from being thrown across the kitchen floor the night before. She grimaced and felt the tears well up, from the pain of the beating and from the happiness of having Debie with her, but she did not want Debie to see her cry, did not want to admit to the hurt, and so kept her tears of bad and good things inside, holding Debie tightly.

In the beginning, Debie flourished. She enrolled at St. Pascal Baylon Junior High in Oakland. Always an average student, she excelled at sports. Swimming was her favorite; she had been a natural swimmer from the age of six, just took to it, nobody knew why, and she quickly became a key member of the St. Pascal team. Her bedroom was soon festooned with blue, yellow, and red ribbons that she won in competitions with other schools. At home, there was always a party going on, always a strange parade of exotic characters dropping by. They would arrive two or three at a time, often on Harleys, big guys with lots of hair and tattoos, walk in through the open screen door and say, "Where's Herb? Herb here? Is Herb around?" Herb was always around; he rarely worked anymore, just stayed at home and had fun—or so it seemed to Debie. Herb would come out of the bedroom with little Baggies or carefully folded squares of foil or small vials—it took Debie a while to figure out what was in them and when she did, she didn't make much of it—and greet his friends: "How ya doin', Fuckface. Hey, Chief. Cretin, how's it goin'?" He would hand them one of his little parcels and they would give him money and either leave right away or disappear into the bathroom for a few minutes and then come out and be in a mood to party all night. They were so much more fun than Clarence, Debie thought, thinking of how he never kept even one can of beer in the house, made the kids drink vegetable juice from his Champion juicer every day, never let them forget that he was the first guy in Sonora with such an appliance, and if you didn't have your health, then what did you have? For the first time in her life, Debie started to feel like maybe she counted; sometimes Herb asked her to help him out, sending her around the corner to pick up a Baggie from a friend, or drop a package off with his buddy who worked at a gas station down the street—she was part of a family and she liked it. Herb's friends, bikers all, or red-and-whites as they were known, were nice, always asking how she was doing with her swimming, what she wanted to do when she graduated, things her father never asked. Then one night a biker who called himself Viking appeared on the front porch with a lion tethered to a chain-link leash. Herb invited him inside. "You can pet it," Viking said to Debie. "It won't hurt." Fearless, intuitive, sensitive to animals, Debie stepped carefully toward the jungle animal. The animal seemed old, tired, although the wild glint in its eye still told of furious kills and feeds eons ago, still flashed menace like a sharp blade. Debie let the lion sniff her palm as Viking held the leash taut. She felt his muzzle, then moved her hand along his jaw up to his broad forehead. Viking said it was the first time he heard the lion purr. Debie said that she learned at school that lions don't purr, they roar. Viking told her that whatever she was doing, she had a way with him, and Debie replied that was because she was a Leo.

A few days later, Herb beat up Rose in front of Debie. Rose fought back, hard, but in the end, Herb overpowered her, punching her in the jaw and sending her into the kitchen cabinets. Debie ran into the streets. From that moment on, Herb's friends would give Debie sanctuary; what was between grownups was just the way it was. There was always someone at their home a few blocks away. She loved doing errands for them, making them happy; they were so demonstrative with their affection, unlike her father and stepfather, so she never said no when she was asked to take brown sacks to a local dry cleaner, little brown bags about as heavy as a lunch-meat sandwich or maybe an apple, never thought to look inside until one day, after having made the delivery dozens of times, she just happened to take a peek. The bag was filled with thick wads of cash.

It took twenty years for Rose to tell Herb that she knew he was dealing speed, bennies, and pot from her house. She knew that if she confronted him, she would take a beating, worse than all of the others. The kids were long gone and there would be no one to stop it. So she decided to drive to Sonora and pick up Herb's mother. Maybe Herb would back off if she was supported by the woman who gave him life.

Rose and Hazel Powell waited in the living room for Herb to come home from a bender. Hazel was so small and fragile, especially now in her seventies. As expected, Herb stumbled in late, reeking of whiskey and speed sweat, slightly taken aback at the sight of the two women. He hadn't seen his mother in seven years. Calmly, Rose asked Herb for a divorce. She told him she knew he was selling drugs and she was tired of all the beatings. Herb made a quick move, somehow the violent surges always sobered him up and allowed him to move fast; she tried to run but he grabbed her, punched her in the face, in the ribs, then tackled the china cabinet and hurled it to the floor, and left forever. Hazel ran into the bedroom and cried.

The women did not call the police; women from Sonora never did, no one from Sonora ever turned anyone in, not ever. They took care of things themselves, because that's just the way it was: that was Miner's Law. But Miner's Law is a more complicated code than those outside it realize; it is not just a prescription for revenge. Allegiances are fierce and deep, end just as quickly as a slash to the bone, then begin again when the need arises, when the people who live in its world are heartened—if only for a moment—by their majestic surroundings, and dance once again with each other. Years later, Rose's life took a strange turn. She was living in the one-room railroad flat in Oakland, the only place she could afford on her waitress salary at Dave's.

One night she was standing at her ironing board, pressing the wrinkles out of her blue uniform and listening to a country-and-western record; this she did on many nights, in fact, ironing was a kind of meditation, it calmed her and helped her stop blaming herself for what happened to her kids, especially Debie, who, the last Rose heard, was in with some bad company. The phone rang; it was Herb. He was living with his mother in Sonora. There was a storm, he said. Can you hear it, Rose? He held the phone to the rattling windows. Rose remembered those storms in the Sierras. Sometimes they could shake a giant sequoia to its very roots; maybe sometimes, not very often but sometimes, they could make a man who is old and tired take a look at his past, and then look away in shame. Rose said that yes, she remembered the storms; in fact, she missed them, missed how clean and new everything looked after the storm had moved on, missed the smell of the forest as it dried when the sun came out again. Finally, Herb asked Rose to meet him. How about Art's? he said; it was a steakhouse in Oakland, near Rose, a nice one that Herb liked. Rose had no fight left; she agreed, although she wondered what Herb's angle was. A couple of days later, they sat across from each other in a red leather booth, to all the world old lovers who maybe even had once visited Portofino, pictured so seductively on the wall next to them. Herb ordered steak, his favorite, and Rose, who never liked meat, had a plate of linguini and marinara sauce. They ate in silence for a little while. Finally, Herb asked her how her meal was. Rose smiled a little smile and said it was okay, it had been a long time since she had a really nice meal, and then Herb told her some gossip about people in Sonora. Rose said she had been meaning to get back one of these days, she wanted to visit her mother's grave. While cutting a piece of steak, Herb had a stroke. For the next few months, Rose took care of the man who had tried so many times to kill her.

PART TWO

Land of Plenty

I wish all the ladies were pies on the shelf.
If I was the baker I'd eat 'em all myself.
Left . . . left . . . left, right, left . . .
I wish all the ladies were bells in the tower.
If I was the hunchback I'd bang 'em on the hour.
Singin' hey boppa-ree-ba, hey bobba row . . .
Wish all the ladies were holes in the road.
If I was a dump truck I'd fill 'em with my load.
Left . . . left . . . left, right, left.

Gulf War marching cadence

Who is this coming up from the desert
like a column of smoke,
perfumed with myrrh and incense
made from all the spices of the merchant?
Look! It is Solomon's carriage,
escorted by sixty warriors,
the noblest of Israel,
all of them wearing the sword,
all experienced in battle,
each with his sword at his side,
prepared for the terrors of the night.

Song of Songs 3:6–8

At the top of a hill, Debie McMaster stopped her beat-up Chevy pickup, rolled down the window, surveyed the land, and burst into tears. She had been on the way to Twentynine Palms for years, generations it seemed, and now she was home. The year was 1984. She and her three kids and Corky, the ten-week old family pit bull, had been on the road for days, running on empty, a scratchy old tape of George Strait, and a prayer. She was anxious, skinny, down to her last pair of jeans and her only pair of shoes—her trademark red cowboy boots—relieved and scared all at once. The family had been driving slowly south from the Tehachapi Mountains, taking the 5 to the 99, putting distance between yesterday and today, hooking up with the 15 outside of Barstow, then heading west, finally turning south again on a hilly two-lane called Old Woman Springs Road, pointing the compass for what the grapevine said was the land of redemption. After years of abuse of every kind, years of involvement with the Hell's Angels, at thirty-one Debie McMaster was finished and starting over in the desert. "Here, Mom, have a cigarette," said Mandi, Debie's eight-year-old second daughter. The little girl handed the puppy off to her younger brother Jason so she could reach the glove compartment, fumbled through some papers and matches, and retrieved a half-smoked butt. "Thanks, baby," Debie said, still weeping. She punched in the lighter and lit up, took a nice, long drag, and sighed the deep, sweet sigh of resignation as once again, she surveyed their future. There was nothing here, the slate was blank: home at last.

The population of Twentynine Palms, including the local Marine base, is twenty thousand. Visitors—a mix of well-to-do Los Angeles grunge, Eurotrash, rock climbers, rock bands, members of the Hollywood ruling class—come here for its extreme beauty, a landscape that looks like the place where life began, a big and bottomless cipher filled with white sand, startling forms of cactus, and frogs that manifest after a desert rain. Visitors often stay at the fashionable inn just outside of town, and know little of the lives of the people who clean the rooms, weed the gardens, wash the dishes. Once, a long time ago, such laborers were referred to as lower class. Nowadays the invocation is no longer made—according to surveys and pollsters and all manner of national mood barometers, everyone is flush, everyone feels pretty good; nowadays the people who inhabit the world of minimum wage and below no longer have a name. But sooner or later, they share the same address; sooner or later they make their way to Twentynine Palms.

Except for certain geologic spectacles, Twentynine Palms is the last stop on Highway 62, which runs east from Interstate 10 (old Route 66, starting at the Pacific Ocean) to the Colorado River on the California–Arizona–Nevada border. Here, you can turn left for Laughlin, Nevada—Las Vegas with jet skis and low-level celebrity shows—or right for Kingman, Arizona—Timothy McVeigh territory, outlying chamber of Mojave's vast white heart. The one-way-in, one-way-out blacktop connects the beery, trashed-out, and beautiful hamlets that mirror L.A. just as Munchkinland foretold of the Emerald City. Out in the desert, the mundane takes on mythological proportions; there's so much space that everything looks more like itself, and there's no need to dress anything up. Out here, a 7-Eleven sells life, not just snacks, water, and lottery tickets. Out here, last names don't matter; old friends know each other by first name only, or desert affiliation (Water District Judy), or habit ("He's into Jagermeister"; "She likes to rap with the Samoans"; "He collects cigarette lighters"). Out here, manners are beside the point; the weapon of choice is the knife, purveyor of the direct and immediate message. This is the shadow side of Twentynine Palms, the side that its many visitors do not see; it is a side that lives and flourishes like cowbirds on the castings of the local Marine Corps Air Ground Combat Center. MCAGCC at Twentynine Palms—"makax" in native parlance—is the biggest Marine base in the world, accounting for half of the town's population. It occupies a wide-ranging site of 932 square miles north of Highway 62; here is warehoused a vast supply of live artillery, a cache of destructive, man-made energy that matches but can never outdo the seismic vibrations emanating from the fragile sands below.

The approach to Twentynine Palms is a long Mojave parade route of fraternal lodges, cheap motels, and cross streets with names that beckon—North Star and Lupine and Ocotillo—and front yards with pit bulls that tell you to forget about it. Within its range, the radio station fades from Christian advice shows to the raw and basic words of other evangelicals with beards, primarily ZZ Top. Behind the distant calls of songbirds and ravens on endless garbage runs, the music of the Mojave is 1970s all the way—Foreigner, Ozzy, Bad Company—the stuff that sounds best on big, old speakers in desert taverns where people get their mail, cash their paychecks, and suck with great purpose on the last of their generic cigarettes. As the altitude hits about three thousand feet, there is an ever-so-slight change of atmosphere: the trail markers on Highway 62 speak no longer of gluttony (SCAMPI, SCAMPI, SCAMPI—what else but Vegas over the horizon, ever trolling for suckers), but of sin and redemption—24-HOUR BAIL/CALL DAY OR NITE—and then, the edge of Twentynine Palms, a sign announcing, VIRGIN MARY SPEAKS TO AMERICA/DIAL 1-800-882-MARY.

And then comes the Joshua tree, the wildly gesticulating plant that characterizes this town that really has nothing to do with the palms after which it is named. The Joshua tree grows in only one place in the world, and this is at two to six thousand feet in the Mojave (although there is a desert plant that looks just like it in the Middle East, having attained the same features because it adapted to an almost identical environment). Unlike many who traverse the paths of the Mojave, the Joshua tree now has a permanent home—in the Joshua Tree National Park, which comprises 560,000 acres to the south of Highway 62. A haunting domain of peace and quiet and heightened beauty on a north-south parallel with the Marine base, its polar opposite, the park and its signature denizen are a prime source of income for Twentynine Palms, especially after the spring rains when the scent of creosote infuses the air and the desert is a thick carpet of rare flowers. Some are delicate, like the soft, yellow Indian paintbrush (which blooms first because its color attracts bees) or the low-lying blue-and-white desert primrose which, in a very wet year, cover the bajadas as far as the eye can see; some are a little more assertive, like the bushy red chuparosa. And then there is the Joshua tree with its full, white-petaled blooms—a surprise gift from this frenzied plant.

The Joshua tree was named by the westering Mormons who believed that it embodied the spirit of Joshua, beckoning them to the Holy Land. Prior to its endowment as a biblical herald, the Mojave Indians regarded the tree as a stairway to heaven. The plant appears to send or receive messages, depending on the pilgrim's mood. For some, it has a physical function, it is simply a home—desert opportunists such as cactus wrens and lizards stop here and move in, not because they've been invited by previously arranged nesting areas, but simply because there's nowhere else to go. "It is a weird vegetable," Colonel John C. Frémont wrote in 1854, hinting at future military use of the desert, "perhaps the ugliest I've seen." In fact, the Joshua tree is not really a tree at all, nor is it a cactus, although, with its spindly branches and leaves that are sharp as daggers, it looks like both. It is actually a member of the lily family (*Yucca brevifolis*); like a lot of things in the desert, it is living a beautiful, bald-faced lie.

The Marines like to say that no one comes to Twentynine Palms on purpose. But history says otherwise. Like the Joshua tree, those who plant themselves here are also misunderstood, seek and require distance, provide shelter for other creatures but only because they happen to show up, possess an awkward beauty, and will hurt you if they are crowded. Although many locals are not natives, in this sandy soil they can sink roots and flourish, and leave fast and without a trace if required. Ancient land of rebirth, fertile triangle of creosote and ocotillo and sage, this is where American wanderers come to start over, to rest, to hide. (However, in spite of the Old Testament panorama of patience and suffering and nomadic sanctuary, one of the planet's original tribe of wanderers is absent: the local paper once ran a headline which stated, JEWS A MYSTERY IN AREA.)

Long before the town was named, the site was forming itself into an intaglio of gas-food-lodging at the center of migratory trails. Over 200,000 years ago, a massive 8.1 earthquake off the San Andreas created the Pinto Basin Fault. For the next ninety thousand years this fault line erupted in at least nine more major quakes, producing a fine powder under the surface called "fault gauge." Over the centuries, the particles of fault gauge coagulated into a kind of cement which trapped water in the fault zone. Soon the water rose to the surface and began to attract the region's earliest travelers—coyotes, saber-toothed cats, giant ground sloths, mastodons, camels, small horses, antelope, and flocks of migrating birds. About fifteen thousand to twenty thousand years ago, when the ice age ended, the first Native Americans made their way to this spot, a lush oasis shaded by a grove of fan palms, the only source of water in a land that was rapidly desiccating. Unlike Indians in other parts of the country, the Indians who lived here from the time of the ice age until about 1000 A.D. left a simple record of how they hunted and gathered and prepared food, and exactly what it was. There are no mysterious ancient mounds, no accumulation of giant boulders that might have served as an ancient calendar. Even the prehistoric pictographs in the surrounding rocks reveal few clues; for the most part, scientists have yet to interpret their meaning, although it is clear that the early inhabitants of the oasis favored the color red. As these Indians moved on, or perhaps disappeared, others found their way to the oasis. First, there were the Serranos, a branch of the Shoshones, who arrived several hundred years ago. According to Serrano legend, they were wandering in the desert for a long time; one of their creation myths suggests that in the beginning, there was nothing. "There was no sky, no earth, no water, but just empty space. In this empty space there became two clouds. One was called Vacant and the other was called Empty. They were brother and sister." In the 1840s and 1850s, Utes from the east raided cattle ranches in the nearby Cajon Pass, running the Serranos off the oasis. The few who remained were wiped out by a smallpox epidemic which swept through the Indian population of Southern California in the 1860s. Around this time, the Cahuilla Indians passed through the region. One of their myths tells of an evil presence on the land. In a canyon on nearby Mount San Jacinto dwelled a cannibal spirit named Tawkish. "He is always on the lookout for people whom he can steal or whom he can rob of their souls," the legend says. "In his home he feasts upon his human victims." In 1867, a war between the Chemehuevi and Mojave Indians broke out along the Colorado River to the east. Greatly outnumbered, the Chemeheuvis fled westward; some of them settled at the oasis. Like the Indians who lived here before, little about them is known.

There was one Chemehuevi, however, who became famous—not because he was a leader of his tribe, and not for his noble pronouncements. It was because he was a killer, and his name was Willie Boy. It was 1910. Every tribe in the country had been run off its land. In Twentynine Palms, the railroad had bought the land that belonged to the Chemehuevis and they were relocated to a small reservation. On the reservation, as on most others, there was a big drinking problem. Willie Boy was the biggest drinker of them all. He wanted to marry the chief's daughter but the chief said no. So Willie Boy killed him and the pair ran away into the Mojave. And so began the last Indian manhunt, a search for a renegade red man that was hawked on the front pages of every Hearst newspaper in the country. WILLIE BOY KIDNAPPED AN INDIAN VIRGIN, screamed the headlines. WILLIE BOY WAS A DRUNK SAVAGE. How did he get away? For several days the pair eluded a white posse in the rugged territory outside of Twentynine Palms. But the young girl tired of running. So as not to bring shame on Willie Boy and risk capture, she killed herself. When the posse found her body, they believed they had another murder on their hands. The search intensified. FIND HIM!

demanded William Randolph Hearst in thirty-point type. FIND HIM NOW! Finally, they did—riddling the country's last unfettered Indian with too many bullets to count, and shooting his horse in the gut; both bled to death in a grove of Joshuas. Willie Boy's boots and bridle and saddle were shipped to New York for display. LAST INDIAN MANHUNT ENDS! proclaimed the papers. WILLIE BOY CAUGHT! It was a sad finish to an obscure tribe whose few remaining members died in bar fights. Every year, residents of Twentynine Palms saddle up and reenact the search for Willie Boy, retracing the steps of the posse that rounded up a defiant, drunk Indian who just wanted to marry a sweet young thing. More than a celebration of his capture, it is an appreciation of the kind of desperado behavior which produced it, a moment of history which fuels the town's shadow side, a side not depicted on any of the historical murals painted in recent years to promote tourism.

In 1994, in fact, the chamber of commerce officially subtitled Twentynine Palms "An Oasis of Murals." It was a project that the town's merchants and third- and fourth-generation desert gentry had planned for years, convinced that there ought to be something besides nature and the Marine Corps to generate income, hoping that the murals would compete with the scenery and keep tourists on cash-starved Main Street long enough to purchase snacks and souvenirs. On the north wall of 29 Palms Eye Care Clinic is a panorama of Dr. James B. Luckie, a Pasadena physician known as "The Father of Twentynine Palms"; he was the first to send World War I veterans who suffered from the effects of mustard gas to the desert for healing. On the south wall of 29 Palms Gemcrafts is a portrait of Jack Cones, "The Flying Constable," who patrolled the town in his Piper Cub. A mural on the east side of 29 Palms Thrift depicts the Dirty Sock Camp, a turn-of-the-century mining berg that derived its name from a method which used socks to separate mercury from gold. On the south side of the Video Connection, a vast mural portrays Frank and Helen Bagley, among the first of the town's many homesteaders. On the north side of the Bowladium, there is a panorama entitled "Desert Flood," a striking depiction of a washed-out intersection after a surprise cloudburst.

But inconvenient weather is as close to the bone as official history ventures. A more accurate slogan for Twentynine Palms would be, "Don't ask, don't tell." The water still flows out of the Pinto Basin Fault and life continues to flourish around the oasis. But today, as nomads have diversified and fragmented along with the rest of the country, the town is a way station for latter-day blackguards, exiles, and refugees—people between minimum-wage jobs, parolees who are not wanted elsewhere: gang members looking for new turf, prospectors who subscribe to *Buried Treasure* magazine, all-night blackjack players, bikers, hikers, people who talk in tongues, retirees, asthmatics, methamphetamine chefs, welfare mothers, runaway kids, people who are stranded here because their driver's licenses have been revoked after one too many DUI busts, people who are running away from other people, people who are running away from cities, people who rent apartments that are a "walk to McDonald's," people who are well versed in *The Federalist Papers* and the Second Amendment, and when asked to observe the law are quick to repeat every American's first civics lesson: "It's a free country, I can do what I want. Where does it say I can't drive fast?" The intense worship of personal rights is fueled by a variety of weekly drink specials at the local bars; as the old saying goes, "The rich get richer and the poor get drunk."

On beery afternoons in taverns, it's not unusual for a patron or two to claim lineage to Wild Bill Hickok or the Clanton brothers or Jesse James. The stories have the ring of truth, for the tellers are generally not boasting but full of shame, tweaked by a blood legacy of hard-core killers, as if they are related to ghosts, as if they—the flesh and muscle and bone of American mythology—know the real meaning of things. Yes, here live and here have died and here will continue to come the progeny of gunslingers and outlaws and boozers and brawlers who built this country, who once raged across the land, whose blood has quenched and quenched again the desert sands. Their history haunts them, stalks them, makes them edgy even as it makes America get up in the morning and whistle a happy, hollow tune. They are drawn to this elevation, this town of Twentynine Palms, where their eyes can fix on nothing but space, space and believe-it-or-not plants, and they can calm themselves, and try to start all over again.

For Debie McMaster, a lifelong, self-proclaimed worker-bee bartender carrying the résumé of the rootless, Twentynine Palms was the land of opportunity. She was a good barkeep and knew she could get a job in this oasis of pubs. It also seemed like a nice, quiet place to raise kids. And in the desert, there was plenty of solitude; maybe her migraine headaches would finally go away, along with the memory of how two generations of women in her family had died in the streets. Soon, she found plenty of company; she was not the first minimum-wage-slave single mother to start over in the Mojave—the female tumbleweeds of the land were here in great abundance. They had always been here, and in a way, they had always been camp followers. South of Highway 62 in the park, there is a little grave bordered by a flourish of ocotillo. The headstone says, MARIA ELEANOR WHALLON, 1885–1903. A marker nearby explains that the eighteen-year-old girl had come to the Mojave with her mother, who had landed a job as a cook at a mining camp. Maria was ailing and her mother had heard that life in the desert could work miracles. Soon after their arrival, Maria died. Her mother stayed on.

A hundred years later, many of the women who found themselves in Twentynine Palms had attached themselves to the military, singing a tired version of "Where the Boys Are"; the road before them one of taking care of men whose choices in life were not mining for gold but joining the service or flipping burgers and who chose the former, with their women following them from base town to base town. As the bases started to disappear toward century's end and the towns around them faded into oblivion, more and more of these single mothers converged in the few remaining military towns. Members of neither the local establishment nor military wives' clubs, they became the scourge of those in control, viewed simply as women who threw themselves at Marines to take advantage of them by having their babies and thereby becoming eligible for base benefits. Consigned to a life at civilization's edge, they were not unlike certain desert critters, demonized, misunderstood, with lives so fraught with danger and hardship that to guarantee survival, they had lots of babies, banking that maybe one or two would carry the line, range into territory that is hospitable.

Compared with some of the women in town, Debie's brood was small, just three; but she was convinced that if the desert could heal veterans of World War I who had fought in the trenches and inhaled mustard gas, then surely it could work its magic on her family. The Mojave was the one region in California Debie had not yet tried. She had decided high altitude was bad luck; she had not visited the upper elevations in sixteen years, since she had left her father's home to live with her mother in Oakland. She had long ago vowed never to return to Sonora. And she had vowed to distance herself from her mother.

Victorville is two hours north of Twentynine Palms, on winding Old Woman Springs Road, Route 247, the same road that Debie

McMaster had taken a long time ago when coming down from the mountains and heading into the desert with her children and her dog Corky to seek a new life. Like Twentynine Palms, it is a town in the orbit of the military, extreme geological beauty, and ancient Mojave history. In Victorville itself is the defunct George Air Force Base, now a mall of sorts, a vast tract of structures and runways which serve as a quick study of life in the desert: there is a truck-driving school, an aircraft-maintenance terminal, housing for the homeless, a nine-hole golf course, headquarters for the High Desert Domestic Violence Organization, and a dormitory that will become a medium-security prison, pending the expected removal of the Mojave ground squirrel from the threatened and endangered species lists. To the northwest of Victorville is Edwards Air Force Base, where Chuck Yeager summoned up the right stuff and broke the sound barrier above the vast white flatlands outside the town of Mojave. To the north, at the gateway to Death Valley, is the China Lake Naval Weapons Center, which is exactly that, as well as home to Little Petroglyph Canyon, the most extensive and complicated rock art site in the country. (Only recently decoded, it suggests that life among the Native Americans who dwelled here was far more sophisticated than white historians have long believed, depicting a multilayered dream life in which shamans crossed the door between this world and the other and became the animals they hunted.) Little Petroglyph Canyon is a protected site and open to public viewing. Pending the schedule of military maneuvers, visitors must provide their social security numbers to the Navy before entering on guided tours. To the northeast of Victorville is Fort Irwin, a weapons-storage area and source of many civilian jobs in the area. During the Gulf War, army troops studied desert warfare at its rigorous National Training Center. The grounds were such an effective staging area that after the war, commanders agreed that winning back the desert in Kuwait was a simple task compared with warfare in the Mojave. As if to imbue soldiers with the qualities they may need in battle, there is a certain petroglyph on the base at Fort Irwin, one that is not seen elsewhere in the Mojave. It is not romantic or appealing in the way that certain stone drawings of birds and bighorn sheep are; with its open jaw and exposed teeth it is simply menacing and has to do with a particular state of mind desired by the desert warrior. It's a jackal, and as far as anyone knows, the jackal has not ever lived in the dimension of time and space known as the Mojave and yet here it has landed, across the eons, homeless and rootless like those it tracks, sleeping in stone on a military preserve, its spirit waiting to possess the appropriate vessels, its soul waiting for certain wars to be waged, certain ill-fated prey to enter its sights and warm its blood.

Yet unlike Twentynine Palms, this region is very sad and desolate, a screeching, minor-key tone that seems to radiate in all directions. The very history of Victorville has left it in a kind of permanent death throe, permanently drawing its last few breaths, on the receiving end of one too many broken promises. Old Route 66—America's route of flight—used to run right through the heart of downtown, but when Interstate 40 was completed in 1984 and subsumed many parts of 66, it bypassed Victorville, leaving a small community of quaint wood-frame cottages which, except for the Route 66 Museum, have long been deserted and are falling apart. Any day now, say town boosters, rich people from Hollywood are going to invest in this historic neighborhood and fix it up. Any day now, construction is set to begin on a new international airport, it's just a matter of time, what with permits and everything, of course the Palmdale Airport will never work out, what a stupid idea! Any day now, the city is going to clean out the hookers and drug dealers who live in the motels up the street. And maybe we'll get that new jail everybody wants to build out here.

But perhaps Victorville was on the skids before the freeway bypass. Perhaps the sadness and desolation derives from its name, which memorializes J. N. Victor, construction superintendent of the California Southern Railroad in 1888 and '89. The name conjures nothing, in the way that the names of other desert towns hint at scenery and healing, some even at the same time, like Palm Springs. In its ordinariness, it carries only the ordinary, the banal. But a more appealing name may not have presented a different picture; perhaps what came to be known as Victorville was always doomed, the way certain restaurant locations always fail no matter what kind of food they serve or how much commerce continues around them. Scenery-wise, Victorville is on the way to other, more spectacular regions. Joshua trees grow in Victorville, but they do not flourish as they do elsewhere; they are scrawny and very far apart and do not appear to have rooted here happily. Perhaps their existence here was an accident; perhaps birds carried their seeds and dropped them off at the right elevation but at the wrong angle for sunlight. Regardless of migratory path, they are reminders of the quality of life in Victorville—hard, resigned, stuck, nothing in sight but gravel. In fact, Victorville's major industry is the manufacture of cement; beneath the town are large deposits of limestone and granite. The other major source of income is the Mall of Victor Valley, the self-proclaimed largest enclosed regional shopping center between San Bernardino and Las Vegas. This is no inconsequential mall; there are dozens of stores there, including Sears and JCPenney, and the items that these stores provide and make desirable inform the lives of the sixty thousand people who live in Victorville. Planning to get married? Buy solid gold wedding bands on the installment plan! Need a new set of tires? We'll work with you! Would you like a triple frappuccino so you can feel good about spending the rest of the day in the mall? Stop right here! Would you like to find a book and read about the rest of the world? Sorry, we don't have any books here, not that we know of, maybe you could try the library, it's downtown somewhere, if we still have a library, they cut way back on their hours after the last election, anyway you don't want to make that drive, it's not a nice area, why not stay here where there are no surprises?

Interstate 15, also known as the Mojave Freeway as it winds north from Devore through the Cajon Pass and into Barstow, intersects with Green Tree Boulevard to the immediate southeast of the Victorville courthouse. A right turn on Green Tree Boulevard leads directly and quickly to the Best Western Green Tree Inn, a motel which because of its proximity to the courthouse has a deal with the state whereby it houses out-of-town witnesses in criminal trials. Even before it became a Best Western, it was a typical Best Western; it had a pool, a coffee shop, a fancy dimly lit restaurant, smoking and nonsmoking rooms, shag carpeting, air conditioners and heaters that made a lot of noise, machines that sold ice, and employees who had been here since it opened way back when, when it was just the Green Tree. In this category of motel, the most expensive in Victorville, there also was a Holiday Inn, but the Best Western was regarded as the better of the two.

Debie napped after checking in at the Best Western, then awoke and went down to the Jolly Roger for a Jack-and-Coke. She spotted a small, nervous-looking black man who was sitting alone at the far end of the bar. She approached him and asked if he was in town for the murder trial. He said he was and she explained that they had something in common, they were both here because that big black bastard killed her daughter. "Hey, that's just me," Debie said by way of the man's reaction. "I say what I think. And your name is?" The man replied that it was Trenton Draper. Debie knew from information presented in 1991 at the preliminary hearing that Draper had seen Mandi shortly before she was killed. "Wasn't my baby pretty?" she said. Yes, she was very pretty, Draper replied, beginning to fidget. "Wasn't she sweet?" Yes, Draper said, she was a sweet girl. And she was very nice, he added, then looked away and said he needed to go get some rest before tomorrow, it was three hours later where he came from, which was Petersburg, Virginia.

On the morning of the trial, Debie was up in time to watch the sun rise. As its early, sharp fall rays illumined the desert, she put on a bathrobe and went outside. She stood on the cement walkway linking the motel suites of the second story of the rear wing of the Best Western which overlooked Interstate 15. She watched the eighteen-wheelers roll by and the traffic to Vegas and then noticed a bus with bars on the windows and watched it turn off at the overpass which led to the courthouse and realized it was the San Bernardino County Correctional Facility bus which carried Valentine Underwood to trial. Every morning from now on, until the trial ended, she would stand outside her doorway and watch this bus arrive in Victorville, shift its gears down as it turned off the freeway, and chug toward the courthouse, and although she did not like the idea of having to awake and know that the man accused of killing Mandi was passing by on the pavement below, she took a strange kind of comfort in marking time this way, seeing the progression of the trial by way of a bus which arrived every day just as surely—and with more noise—as the court clerk flipped over a new calendar page each morning. With the bus out of sight on this, the first day of the trial, Tuesday, October 14, 1997, Debie went back inside and prepared, plugging in her hot rollers and making a cup of Maxwell House on the Mr. Coffee she had bought at the local Pic 'N Save, a short walk from the Best Western.

Although the murders of Rosalie Ortega and Amanda Lee Scott were two of the most grisly in San Bernardino County history, the Victorville courthouse was not packed with spectators on the day the trial began. The newspaper of Twentynine Palms, *The Desert Trail,* had not sent a reporter and, in fact, had ceased coverage of the case soon after an arrest was made, occasionally running briefs about delays but generally ignoring it until an article appeared in *Los Angeles* magazine in 1996. Short of staff and unable to send a reporter from Twentynine Palms to Victorville every day, they had made an arrangement with the *Victor Valley Daily Press,* which was covering the case, to reprint its coverage. In addition to the reporter from the *Daily Press,* the only other spectators in court on the first day were Debie, Jesse Fulbright, Mrs. Lee Johnson (the wife of the investigator for the defense), and a few courthouse employees who had heard about the long-running case and wandered in and out.

The venue may have been Victorville, and the participants not rich or famous, but everyone had dressed in accordance with the formality of the proceedings, the pomp and circumstance required by the country's most profound civic ritual. Debie arrived in olive-green slacks and matching double-breasted jacket and pumps which her oldest daughter Krisinda had helped her pick out months before for the occasion. Both lawyers—the same height, about five feet four inches—wore nice navy-blue suits and white shirts, cutting beefy if slightly edgy figures who were ready to get down to business. Lee Johnson, the big, rangy, former cop who was the defense investigator, wore his best gray suit. Sergeant Tom Neeley, the San Bernardino County cop who was one of the investigators for the prosecution, wore a nicely cut black suit. The court reporter for the day, Barbie Roelle, wore a brightly colored A-line dress that had lots of bounce and her hair looked freshly styled, done in the layered Farrah Fawcett style that is a desert standard. Rick Alcantara, the bailiff, wore a starched brown uniform and shiny black shoes. The twelve jurors had clearly chosen the day's wardrobe with much thought as well. Most among the jury (five women, seven men; two blacks, three Latinos, seven whites) wore dark colors; they looked scrubbed, brushed, buffed, ready to take on what the judge had warned them during jury selection was likely to be a two-month trial in which much gruesome evidence would be presented. "I promise we'll be finished up by Christmas," the judge had told them on that October day. Judge Rufus T. Yent looked every bit the Superior Court judge on opening day of a major murder trial, cutting the figure of a kind of desert wise man: his wire-rim glasses and thinning silver hair and perfect silver goatee and mustache were just the thing for a flowing black robe, and for the interpretation of statute.

And then there was the man who allegedly had caused everyone to dress and act their most proper, and convene at a courthouse in the middle of the Mojave Desert: Valentine Underwood. No one had sent him a new suit for his trial, nor could he present himself in a Marine uniform—since his arrest, the Corps had considered him AWOL. His attire was the orange prison jumpsuit worn by all of the high-security San Bernardino County incarcerated, and his accessories consisted of ankle chains.

Prosecutor Gary Bailey rose to make his opening statement. With enlarged photographs of the victims leaning against a blackboard at the head of the courtroom—a ninth-grade class photo of Mandi, with her easy smile and baby fat, still a child, and there next to her was a joyful Rosalie, in a sundress, holding her black hair up over head, posing like a model—Bailey said that a bloody handprint on a wall, a blood drop on a magazine, and a partial shoe print on a knife linked Valentine Underwood to the murders of the two friends. The state would prove beyond a reasonable doubt that on August 2, 1991, between 9:15 and 10:45 A.M., the defendant raped and murdered each victim. Rosie and Mandi had been stabbed dozens of times in the chest and neck, Bailey told the rapt jury. Scott's hands were tied behind her back with a telephone cord, and Ortega, who was not bound, had defensive wounds to her arms. Bailey asked Sergeant Neeley to help him display a blown-up map of the crime scene. "Amanda—Mandi—is right here at position fourteen," he indicated. "Rosalie is here at position twelve. The knife, a ginsu with no hilt, was found right here," he said, pointing to a location on the floor of the living room. "In a few days, I will introduce the knife. Both girls had wounds on the left side of the neck. The sex kit tests revealed evidence of sexual assault on both girls. In position twenty, on the wall about four to five feet above Mandi's body, investigators found a bloody palm print. That wall will come in as evidence. In position fifteen, there was a magazine with a drop of blood near Mandi. In position seventeen, there was a trash can with a bloody white towel . . . Everything matches. The towel contains a mix of Mandi's and Underwood's blood. The knife has a partial shoe print and at position eighteen, there is a pink comforter which was found to have a footprint, both of which match the shoes the defendant was wearing at the time of the murders. His shoes tested positive for his own blood. And finally, we have DNA evidence, which also links the defendant to the murders. I know scientific evidence can be boring and may be hard to understand. But I think you'll find our DNA evidence very exciting. DNA is the blueprint of life."

Over the weekend, after Bailey's opening statement had concluded, other witnesses arrived at the Best Western. Jessielyn Gonzalez, Rosalie Ortega's sister, drove up from Camp Pendleton, where she and her husband were stationed. Debie had not seen her since Mandi's funeral and felt uneasy about having to spend the next couple of months with a woman with whom she shared a terrible bond. Jessielyn called Debie from her room and asked if she could come over. Debie lit up a cigarette and said sure, although she really wasn't sure she meant it. On her way to meet Debie, Jessielyn stopped at the front desk and, as Debie had done earlier, left a message for Timothy Carmichael to please call her in Debie's room if there was no answer in hers. Unlike Debie, Jessielyn knew Carmichael, and liked him; he was the first nonabusive boyfriend Rosalie had had and Jessielyn was convinced that Rosalie and Timothy would have married as soon as Carmichael finished up his stint in the Corps. She hadn't seen Carmichael since the preliminary hearing when he had testified about discovering the bodies of Rosie and Mandi, and although they had not kept in touch, and she had not known him very well, she spoke of him as a very close friend and was looking forward to a reunion, however grim the circumstances, with a man who had treated Rosalie with kindness. Jessielyn stubbed out her cigarette on the cement steps outside Debie's room, not knowing if Debie smoked, and raised her hand to knock on the door. But Debie opened it before she had the chance. Without hesitation, the two women embraced, quickly, and then backed away from each other, not knowing what else to do or say. Jessielyn said that she didn't know how she was going to get through the trial, and burst into tears. Debie said that she would help her: "That bastard's never gonna see me cry." The phone rang. It was Timothy Carmichael. "For you," Debie said, handing it to Jessielyn. Debie overheard the conversation and called from across the room for Carmichael to come on up. "We'll have a witness party," she said. In a few minutes, he knocked on the door. Debie opened it and introduced herself. Again came an intense but quick embrace, Debie cutting it off first, overcome with feelings that had to do with the fact that having discovered the bodies, Carmichael knew, would always carry with him, the horror of the crime scene, the way Mandi looked at the end, something that even Debie herself had not been permitted to experience.

It was sad to know that Carmichael had discovered the murders. Although an ex-Marine, which meant that he, like all

Marines, had been trained as a sniper, he was one of those ever-innocent-looking souls, of lanky frame and medium height, with sweet eyes and one of those smiles that makes everyone feel good. From a poor black family in Detroit, he had joined the Marine Corps in 1990 as a way out of the projects. He was grateful for the ride, and it showed. He and Jessielyn hugged tightly and Jessielyn started to cry. "Jessielyn," Debie said, "I told you, you can't let motherfucking Underwood get to you. If you do, you're never gonna get through this trial." Jessielyn wiped her tears and smiled through them and said she'd try not to cry. The three of them sat down on the big Best Western bed and started to talk. Carmichael needed to talk to Jessielyn and Debie as much as they needed to talk to him. He had been carrying this terrible thing inside him ever since he found the bodies of Mandi and Rosie six years ago in Rosie's apartment. Like everyone else in their lives, he felt if only he had done things differently; in his case, the specifics of it were, if only I had told Rosie I wanted to marry her earlier in the week; if only I had said I love you when we talked on the phone earlier that day; if only they hadn't had that fight two weeks before. "I know she loved you," Jessielyn said. Carmichael said that he knew . . . Yet, he had moved on. Now out of the Corps, married, and the father of two preschool girls, he was living in Tampa, Florida, and working at a Toys "R" Us. He showed photos of his wife and children. Everyone looked very happy; it was a pretty family, and Jessielyn told him so. Then he was overcome by a wave of sadness and said he had been having nightmares about the bodies for the past six years. He had a theory that something evil had happened in the apartment that night, that Underwood was possessed by a demon. He made reference to the number thirty-three, so prevalent at the crime scene. "Thirty-three times two is sixty-six," he said, "and you know what six-six-six is." He added that bad things had happened to two of the other guys who were at Rosie's that night. One Marine who had been at Rosie's playing cards was in prison for murder. "He killed someone at a meat-packing plant in Virginia," he said. "Gwatney Meat Packing." And another Marine who played cards with Mandi and Rosie on their last night was also in jail. "In Missouri, I think," Carmichael said. "Manslaughter." There was a moment of silence and then he roused himself and said, "Fishhead was good people." "Fishhead" was the nickname some of the guys gave Rosie after she cooked for them one night, serving fish heads and rice. Jessielyn hated this nickname, pointing out that it was derogatory to Filipinos. The nickname was to come up often in courtroom testimony. Carmichael patted her on the hand. "We didn't mean nothin' by it," he said. "Just that we liked her. Rosie liked it, too. Sometimes I called her 'Fish.' " Jessielyn took some photos from her wallet. They were of Shanelle, Rosie's little girl, now ten. "Wow," Carmichael said. "She looks just like Rosie." Jessielyn said that Shanelle was living in Las Vegas with Rosie's mother and her stepfather. She gets straight A's in school, she said, and then, turning to Debie, she said that Shanelle's best friend was named Amanda.

In 1970, as soon as Debie had graduated from Alameda High School near Oakland, she left and headed south for Camp Pendleton, the Marine base in Oceanside, California, where her brother Marshall was stationed. Marshall had just returned from Okinawa with a Japanese wife. Could Debie help her learn English? Marshall wondered. He'd pay her for the lessons. All she needed was a plane ticket, Debie said; she'd take care of the rest. On her first day in town, Debie asked Marshall where the Marines went for breakfast. An hour later, she walked into a coffee shop in full black biker leathers and red snakeskin cowboy boots, a Harley logo emblazoned on the back of her fringed vest.

The coffee shop was a dive eatery favored by Marines and sailors and physical laborers in the military town of O'side. A big "Semper Fi" greeting hung in the front window. Inside, the walls above the lunch counter were covered with posters of famous Marine battles. The boys were in this day, lining the counter; in fact, there was not an empty seat in the house. As Debie had expected, the moment she entered and made eye contact, every Marine in the place, each and every janitor, street sweeper, bus driver, every mailman had offered his seat. Not missing a beat, she gave the joint the once-over and then said to the guy wearing the name tag that said BOB, "Hey, Bob, ya know what's wrong with this place?" "Didn't think there was nothing wrong with this place," Bob said. "I don't work here is what's wrong," Debie explained. The platoon cracked up and Bob cracked up and then he said, "You do now, darlin', come back tomorrow at five A.M." As she headed out, he added, "I don't need to tell you what to wear, do I?" "Not on the first date," Debie said. On her way back to Marshall's modest one-bedroom house on the base, Debie saw a sign outside the Playgirl Lounge that said, DANCERS WANTED. She walked in, surveyed the joint, headed for the counter, and asked for the boss. She asked the man introduced as Al if he knew what was wrong with his place. Al asked her to explain. Debie said that what Al needed to do was put her in one of the go-go cages. Al told her she could start the next night, five dollars an hour plus tips. Back at Marshall's, Debie asked for an advance on the English lessons and rented a one-room apartment over a garage on the funky side of Oceanside, which, being a military town, was mostly funky, even on the side of town where military personnel with high salaries lived.

The following day, Debie began her career at the coffeeshop. At exactly noon, a big, handsome Navy midshipman named Carl Fuselier walked in and took the far seat at the counter. As Debie poured him a cup of coffee, he read her name tag and noted the unusual spelling. "I'm an unusual girl," Debie said. "Don't do nothin' like nobody else." Carl ordered a tuna on toast with lettuce and tomatoes. When Debie brought the order, he placed his hand on top of hers before she could take it off the counter, and said he, too, was unusual. If Debie spent the night with him, he said, she would undoubtedly agree. "That ain't gonna happen," Debie said, convinced that this was the man she was going to live with the rest of her life. Carl asked to see her that afternoon. "Can't," Debie said. "After I leave here I got some English lessons I gotta give to my brother's wife." Carl was surprised; Debie didn't look like a schoolmarm. "Told ya, I'm un-u-su-all," Debie said. When Carl finished and paid up and got ready to leave, Debie told him to stop at the Playgirl Lounge later and see her dance. Carl winked and said he was busy. The next day, he came by Bob's, at exactly noon, and every day after that. Every day, Debie had a tuna on toast with lettuce and tomatoes waiting for him at his counter seat. At the end of the first week, Debie was spending every night with Carl. Pretty soon, she was pregnant. Debie wondered why a pregnant girl couldn't be seen dancing. "People will pay good money for it," she said. She was right. She danced at the Playgirl until she was seven months. Then she and Carl got married at city hall.

"Hi, Mom," Debie said on the phone almost two years later, anxiously sucking on a joint and holding back tears. "I need a place to stay." She said Carl had been beating her up since she had come home from the hospital with her first child. Named Krisinda after the infant daughter of her favorite television actors, Christopher George and Lynda Day George, the baby triggered a rage in Carl that got worse with every little infant thing, from needing a diaper change to crying to reaching out just for a cuddle. After every beating, Debie smoked a joint, good stuff some of the Angels were sending her from Oakland.

It blunted her pain, her feelings, helped her pretend that her life was different from her mother's.

One day, after returning from a walk with Kristy, Debie found Carl in bed with another woman. Not missing a beat, Carl greeted her with a big smile and introduced her to his companion. Debie grabbed a few things from the closet—her leather jacket that she wore when riding with the red-and-whites, her red snakeskin boots, a box of baby photos, a bottle of formula—and left. Too embarrassed to go to her brother's and not wanting to call her mother either, she did so anyway, from a booth nearby. "Ma," she said, "I gotta get outta here." "Carl?" her mother asked, noting the tone in her daughter's voice, the tone she had heard so many times before when Debie had called to say that Carl had hit her, what should she do? Debie told her mother what happened, and Rose said she would wire her $200 so she could catch the next flight out of there and come back to Oakland. Debie headed for the bus station, wheeling Kristy up and down the streets of Oceanside until she got to the terminal two hours later, exhausted. There she would wait for the bus to San Diego. In the ladies' room, she took off her sunglasses and examined her reflection; it was as she figured—a black eye, really bad this time. Carl had really done a number on her when he hit her the other day; her eye was more yellow than usual, and oh, shit, her lip was cracked—she looked just like the bum in the next stall, some old bag who had been on the streets for a long time, probably lived right here at the Greyhound. At least Kristy was sleeping now, would never have a conscious memory of this evening, would never remember inhaling the acrid vapors of bus-station disinfectant. Debie lit up the remnants of her last joint, cupping it close to her to hide the act, took a couple of hits, then put some pink lipstick on. She didn't really like looking at herself and wished there was some kind of makeup to cover up the bruises on her arms. She fluffed up her tangled hair quickly while turning away from the mirror, put her sunglasses back on, then pushed her daughter out of the ladies' room, another traveler on the skids trying to carry on. She found a bench in a remote corner of the station, wanting to do nothing but sleep for a very long time, but not able to, even for a minute, because she did not want to let Kristy out of her sight amid all the Marines and sailors and hippies and derelicts passing through the station. Every hour she popped some quarters into the coffee machine so she could stay awake, staring bleary-eyed through her black shades at the departure board way down the terminal, trying to will the bus for San Diego to pull into the station. Several hours later, mother and daughter boarded the bus. Debie unstrapped Kristy from her stroller, folded it up and put it in an overhead compartment, put her baby on her lap, held her tight, and fell asleep. She slept fitfully, dreaming of nothing. In San Diego, she got on a plane with her daughter and headed to her mother's house in Oakland.

"Girls say yes to boys who say no! Girls say yes to boys who say no!" The chant went up from the crowd on Telegraph Avenue in Berkeley. It was 1971. Jane Fonda was addressing the antiwar protesters who had converged to express rage over American involvement in Vietnam. She was dressed like Che Guevara minus the artillery belt and behind her flew the flag of North Vietnam. "We have the power to stop this illegal war!" Fonda shouted to the throng. "We can do it today!" Debie and her brothers stood at the back of the mob. "You bitch!" Debie called out over the cheers and tambourines and drums and whistles. "Traitor!" Now spending most of her time with the red-and-whites since her return a year and a half earlier, Debie was in her biker leathers. Both of her brothers had recently come home from a tour of duty in Vietnam and were wearing standard military issue; they wondered why a lot of the guys in the crowd were dressed like soldiers if they hated war so much. Marshall was home between his second and third tours. Darrel was back from his first—and last: he took an explosive in the gut while guarding an airstrip at Anhoa, was taken for dead, and thrown atop a truckbed piled with Marine corpses. On the way to the body bags, he regained consciousness, was shipped to a hospital in Okinawa, recovered, and was sent home, honorably discharged, with a Purple Heart. Neither Marshall nor Darrel liked to talk about their experiences in Vietnam, in battle; the most they would say was that they were proud to be part of the Marine Corps, just doing their duty for their country, it would not occur to them to do otherwise. As the crowd's chanting got louder, angrier, Debie confronted a long-hair. "You got something against Vietnam vets?" The guy tried to shake Debie off; he was antiviolent, he explained, he was just exercising his freedom of speech. "You see my brother here? He got his guts blown out in 'Nam so you could stand here and talk trash." The long-hair said that he didn't want any trouble and then Debie told him that well then, he shouldn't have crossed her path. As her brothers took her away from the brewing fight, she called out, "Fuck off, you sniveling little pussy! Fuck you and all your snooty-ass hippie fag friends . . ." A couple of other long-hairs gave in to Debie's taunts, called her an ugly bitch, and Marshall threw a punch. A small version of war with six or seven warriors broke out, the McMaster brothers again on the front line of a battle they felt obliged to fight, a battle for honor and respect, a social battle which the family would wage until the end of time, from the barroom to the bedroom to the courtroom.

A few weeks later, another shipment of Americans was heading for Vietnam. The night before they shipped out, Debie and some girlfriends went to the NCO club on Treasure Island, site of a Navy base in San Francisco, to send them off. In addition to her brothers, she knew several members of the military, bikers mostly, guys who might have become career military if the red-and-whites hadn't gotten to them first. Raised in a region whose scripture was the Bill of Rights, with particular emphasis on the Second Amendment, in order to defend all the others, she never questioned the necessity of supporting Americans at war. This fact of Debie's life had now become her calling card, and it promised defiance to the privileged who seemed to despise those who did the country's dirty work, and loyalty to the military, which was under siege at home. This was nothing new, Debie realized, but now it was on the news every night. Poor people were the enemy within, spat on when they came home from Vietnam, while rich people got draft deferments. "I want our boys to leave here with a smile on their face," Debie told her friends as they entered the club in short tight dresses and high heels. The place was crowded with Corpsmen and sailors and girlfriends and young wives, rowdy with the sounds of a live country band, the Seven-Second Men, which referred to the number of seconds a bull rider had to stay on in order to win a competition. Debie ordered a double tequila and reached into her purse. A boy, no more than seventeen, stopped her, put a twenty on the bar, and ordered another. They toasted and downed their drinks and headed to the dance floor for a country swing. "Oooh, you're really good," Debie said as the boy twirled her across the floor; he wasn't, really, but she made him feel like he was, even while he was tripping over his own feet, embarrassed at his clumsiness. This caught the attention of some other boys, and they began cutting in, each one after the previous one had a turn around the floor, the next one waiting for her with a drink. "Oooh," Debie said to each partner, "you're good. You're gonna make me real sad when you go." Before they took a break, the Seven-Second Men dedicated a number to the "pretty lady in the red high heels" and Debie raised her shot glass in a toast to the band. A good-looking Marine, about twenty-five, over six feet, made his way through the swarm of suitors, put his arm around Debie, and said that he'd ask her to dance, but he didn't want to take a number. For the rest of the evening, the pair was inseparable, two-stepping to each other's rhythm as if they had mated for life.

Debie's partner was Maxey Leon Scott—Max to his friends—a good old boy from the oil fields of Midland, Texas. He and Debie began spending all their time together, and a couple of months later when Max shipped out on the *Blueridge* for another Navy tour, Debie was pregnant. He returned six months later and married her at city hall. Amanda Lee Scott was born on August 4, 1975. She was a big baby, eight and a half pounds, and right away indicated that she was happy to be here. She hardly ever cried, and Max would hold her while he two-stepped around the house, cooing that she was Daddy's little girl. Nine months later, Debie had the couple's second child, Jason. Several days later, as the war in Vietnam was ending, Max was ordered back to help evacuate Americans and high-ranking South Vietnamese from Saigon as it fell in defeat. This trip to Vietnam had an especially ominous feeling about it; for the first time since she had known him, Debie felt that something terrible was about to happen, that things would never be the same between them when Max returned—if he returned. "I didn't want no more kids after Carl," Debie said to Max that night. "But I'm glad I changed my mind." Max asked her if she thought he was turning out to be a good daddy as they danced arm in arm around their small dwelling on the base to the tunes of Bob Wills and the Texas Playboys. "Damn straight," Debie said.

Max returned from Vietnam months after Jason was born. Debie was right: something had changed. Since war's end, Max had turned mean. Having been away at war for Jason's first year, he found it difficult to connect with his son; even though Jason looked just like him, somehow his little body suggested that one day he would be shopping at the big-and-tall, but over time there emerged one thing that said he was his mother's boy—he had those cat's eyes, like Mandi, like Kristy, like Debie. The blood tie simply could not overpower other forces. By now Debie had noticed the pattern: Max was not able to decompress from his time at sea, to adjust from months of being away in an undeclared, unpopular war to suddenly having to assume another role in the civilian world, a role that involved things not taught by the military, such as how to be a father to a child he did not know. Max simply did not take to Jason, did not respond to the magic, the flashing dark and light in his eyes, his sweetness, simply did not spark to the son he had always wanted. And he seemed to resent that the *Blueridge* was now in dry dock, take it personally somehow. "Man, you should have seen all the bars of gold the gooks brought in," he said. He couldn't get over it, kept talking about the bars of gold that some of the elite South Vietnamese had managed to take out of Saigon as they fled. He was drinking heavily, much more than before this last tour, several six-packs of Bud every evening and then on to the hard stuff. Debie was concerned and called the red-and-whites. Since she and Max were now living at the Balboa Marine Base in San Diego, they put her in touch with Joe Evans, the president of the local chapter of the Hell's Angels. The two met at a biker enclave outside of town. Debie felt comfortable, as if she had gone home. She especially liked being around Joe and his wife Kat. They had a six-year-old boy named Ray, whom they had named after Fat Ray, a biker who was killed by a rival gang called the Mongols. Joe and Kat and Ray seemed like a real family; there was a legacy that they were nurturing, passing on, and nothing could tear them apart. Soon she asked Joe if he would be her child's godfather. "My mother said she'll take Kristy if she has to," Debie said, "and maybe you could take Mandi and Jason. That way, if something happens to me, I'll know everything will be fine." Joe said he was honored, in fact if she wanted someone to "talk to" Max right now, it could be arranged. Debie still loved her Marine, and she declined.

The really serious trouble began during football season. The Cowboys were playing the Raiders. Like a lot of Texans, Max was a hard-core Cowboys fan, really believed that they were "America's team"; as the Cowboys went, so went the country. He exploded ecstatically with every completed pass, every down, died with every fumble, every bad call. Debie was a hard-core Raiders fan, like a lot of Northern Californians, like a lot of non-Texans who did not identify with their own teams, who had a boss and didn't like him, who preferred the scrappy street-fighter image of the team with the pirate logo and the slogan "pursuit of excellence." Debie and Max were watching the game. The children were playing in their bedroom. It was a good game, tied in the fourth quarter with fifteen seconds to go. Since the beginning of the game, Max had polished off a couple of six-packs. The Raiders' infamous George Blanda faked a pass to his receiver, then lateraled to a guard, who outran two line-backers and a safety, completing the run for the winning touch-down. Debie jumped up and told Max to settle on a bet they had made. But Max was seething, as often happened when he drank too much, and he lunged at Debie. Debie grabbed one of his size-thirteen boots and threw it at him; he dodged and it hit a door, making a deep hole in the thin wood used for base housing. Max came after Debie again. She ran into the bathroom and locked the door. He ran after her, kicked the door down, lunged at her, and hauled her out. In the hallway Debie grabbed a soup tureen and broke it over his head. He fell to the floor and passed out.

Over the following weeks, Max would drink more and the fights would escalate, triggered not just by a Cowboys' loss but by anything that was not right—dinner was too hot, too cold, Debie was not paying enough attention to him. Debie called her mother and cried. She called the red-and-whites. "Should I send someone over?" Joe asked her. "No," Debie told him, "just needed to dial 1-800-boo-hoo. Sorry about the whining and sniveling. I don't want to be a burden." The beatings became more brutal. Max broke Debie's jaw. Debie clawed up his back pretty good. Max hit Kristy after she mussed her pants. In retaliation, Debie kicked him in the groin. He punched her in the face and she went skidding across the kitchen floor. This time she didn't get up. As he left the house, Max told her that he was leaving her for his high-school sweetheart back in Texas and hoped she and the kids were gone by the time he got back. He had already hired a lawyer.

Four-year-old Kristy stood over her mother and wailed and wailed. Mandi and Jason lay in their beds, drenched in tears. Debie lay still for a long time, coiled up infantlike, with the look of a baby just before it drifted to sleep. For during these few minutes on the floor, she had made a decision: when she got up, she was going to kill herself. She reached out her hand to Kristy. "It's okay," she said. "Mommy's okay." Kristy stopped crying and Debie raised herself off the floor. She went to the cupboard and took out a box of Trix, Kristy's favorite cereal, and fixed her a bowl. Kristy sat at the table to eat. Then Debie tended to Jason and Mandi, changing their diapers and calming them with bottles of warm milk. With the children settled down, she entered the bathroom and shut the door. This was as good a place as any to end it, she thought as her eyes took in the mean fixtures of Marine-base housing, right here next to the toilet: the cheap gray linoleum floor, the towel rack that was too narrow for even the cheapest of bath towels, the leaky shower faucet, the faded yellow of the tiles—how many soldiers had washed up in here after beating on their families? Debie stared at herself in the mirror. Her cat's eyes looked dead—no flashing or twinkling. Time to go. She opened the medicine cabinet and popped open a vial of muscle relaxers. She looked inside and saw that there were more than enough to finish her off. Come on, Debie, she told herself. You can do this. This is nothing compared to all those punches you took. This is a piece of cake. She turned on the faucet, put her head back, and poured the pills down her throat. Then she gulped them down and chased them with water that she cupped with her hands. In a few seconds, she sank to the floor. Her chin fell against her shoulder, her eyelids fluttered and then fell heavy. She passed out, a tangle of flesh and bone on the cold bathroom tile. A little while later, Max returned. Kristy was trying to push in the bathroom door, calling "Mommy," but Debie was wedged against it from the inside. "What the hell?" Max said, barging in, and then just: "Jeez." Debie lay unconscious. The water was still running. A damp pink washcloth halfway covered her face; some part of her had roused during the suicide attempt and tried to apply comfort, had grabbed a life raft, as the other was diving for the lower depths.

No one came to visit her in the hospital during the one week that she floated between now and then. On the seventh day, she took a taxi home. She gave the driver a few crumpled dollar bills as he pulled into her driveway, and took a small measure of comfort when she noted the Harley outside the garage: the Angels were taking care of the kids, at least. Fuck Max—here was a family you could count on when some asshole rocked your world. "No thanks," she said as the driver noticed that she was so weak that she could barely open the door of the cab. "I'll get it myself." She leaned into the door with her shoulder, forced the old thing ajar, and weakly stepped out. "See ya," she said, and headed home. She paused at the front door, heard the sounds of a TV show, the theme song from *Bonanza,* took a couple of deep breaths, then knocked. No answer. She knocked again, with more strength, over the sweeping chords beckoning wanderers to the Ponderosa. The door opened: it was a friend she knew as Crash, a big biker who was out on parole after having served three years for a felony. "Hugs," he said, arms wide open. Debie, skinnier than ever, collapsed into his big arms, all buff and cut from working out in prison, and for the first time in a long time, felt that she belonged somewhere. "Everything's gonna be all right now," Crash said, and Debie just breathed quietly into his flesh. "Where're the kids?" she said. "They're sleeping," Crash told her. "Don't worry. I've been keepin' 'em happy with silver dollars. One for every day that you've been gone." A few weeks later, Debie put her furniture in a base storage unit. She and the kids moved to a small apartment in a poor San Diego neighborhood. When settled, she called to arrange a pickup for her belongings. The attendant informed her that Mrs. Scott had already called and had everything shipped to Midland. "Motherfucking Max," Debie told the Marine. "Ma'am?" he replied. Debie asked him why he didn't require identification when a person made such a request. "Ma'am," he said, "you'll have to take that up with the duty officer of the day." In no condition to tangle with the bureaucracy of the Marine Corps, and once again with Max Scott, Debie gave up on her furniture. It wouldn't matter in this neighborhood, anyway, she figured. She would hardly be the only resident without a nice sofa and dining table with matching chairs. In fact, it wasn't long before Debie fit right in. As the head of a household with an income below the poverty line, she applied for federal aid. To her never-ending embarrassment, she had become a welfare mother.

A year later, Debie had a new address. It was the Motel 6 in Folsom, California. In fact, her last several addresses had been Motel 6. There was a new man in her life—a biker named Monster. He had been busted for armed robbery and received a long mandatory sentence. First he was sent to the penitentiary in Chino. Debie, still collecting welfare and awaiting overdue child-support checks, dropped Kristy off with her mother in Oakland and rented a Motel 6 kitchenette near the prison. A few months later, Monster was transferred to Lodi. Debie and Mandi and Jason followed. Finally, Monster was moved to Folsom. Once again, Debie and her kids trailed him to his new home, and theirs.

If ever there was a snapshot of the caboose of the American train, the Motel 6 in Folsom was it. There was a skinny Vietnam vet with Samson hair and a protruding rib cage. He had lived in the motel for years. He was looking for gold in the nearby hills, no one knew exactly where, and he would never say, even took a different route each day in his pickup to get to his secret spot in case anyone was following him. There was a hard-drinking, robust sixty-five-year-old man who had once been a star on *Roller Derby*. Now he was living on a pension and what little remained of a small inheritance from his mother; "The Heir of Motel 6," he was called by his fellows. There was a biker who loved to bake bread in his electric bread-baking machine, didn't eat much else, just beer, and bread. In the unit next to Debie was a former Vegas showgirl, a wino, with twelve wigs and a pregnant pit bull. Debie and her kids were one of five welfare families at the Motel 6. The children were enrolled at the local school. Sometimes the moms would hang out in one of the units, wherever there was a door open, drawn by the female pulse, some biological urge to state the facts of their lives. They knew that their lives would never be anything but these facts and they found comfort in the recitation. "My ex–old man never sends the child support," someone would say. "I don't even know where mine is," another would say. "Mine refused to believe the kid was his," would come more testimony. "I said let's take a test and he divorced me." They all chain-smoked, and even if you didn't see them smoking, you could hear the smoking, the anger, the disappointment in their laughter, which would rumble first out of their lungs, then from the lower depths, up through their feet and legs from deep down inside the guts of the planet, as if the very womb of earth itself were writhing forth a big belly laugh, heaving up a huge guffaw like the fat lady in the circus who doubled up on cue and then couldn't stop laughing because when she did the circus would be over, the planet would stop spinning in its tracks, and then what would all the men do?

Every community has its class system, and compared with some of the other residents at the Motel 6, Debie was well-to-do. In addition to her welfare income she made money as a bartender and small-time dealer of speed. Her biker friends supplied her with enough to set herself up and keep her business going, asking in return that Debie smuggle the stuff into Folsom to supply their brothers on the inside. The extra income wasn't much, but a few hundred here and there—it meant that sometimes Debie could add meat to the macaroni and cheese she cooked on the hot plate every night, she could make sure there was a fresh head of iceberg lettuce in the mini-fridge, and her kids didn't have to get on the school bus with threadbare clothes like some of the others who lived here. Sometimes, to help out some of the other mothers, Debie would front them an eight-ball or a quarter of speed just so they could pay the rent, get back on their feet after having to spend their last few bucks on bail for a son who was popped for vandalism or shoplifting.

Three times a week and twice on weekends, when visits were allowed, Debie would pay a call on Monster. Sometimes she would have methamphetamine hidden on or in her body, in a tampon for instance, or under layers of hard acrylic nail polish. Debie would leave the conveyors behind, in the ladies' room perhaps, and then the chain would begin: a designated person would find what she had left, extract the meth, pass it to the next person, until it finally reached Monster. "Hello, sunshine," Monster said to Debie on the phone from behind the glass screen. "Hey, baby," Debie said. "How ya doin'?" "Fine," Monster said, "now that you're here. I'm always fine when you're here. How's the kids?" "Mandi's learning to skate," Debie said. "And you know Jason. He goes where Mandi goes." They exchanged conversation about their friends: Ghost is back on the street again but who knows how long that will last, how come no one ever talks about the good things we do like Toys for Tots, Sonny was busted again, did you hear what the feds are doing with the RICO act? And then the game was over and like all the other inmates and their visitors during visiting hours, they touched palms through the plate-glass screen and departed for their respective distant corners until legal contact started again.

One night, close to midnight, there was a knock on Debie's door. Mandi and Jason were asleep in their shared bed, and Debie was in her single, catching up on her *Soap Opera Digest* with a reading light, polishing off the last of a butt, coming down from a hit of speed a few hours ago. Johnny Carson in black-and-white with the sound turned low did his golf swing on the TV table at the head of her bed. Debie stubbed out the cigarette, exhaled the dregs, and tiptoed in her oversized Raiders T-shirt to the door, quietly opening it. Like a distant mantra, the voice of Johnny and the responses of his singularly American disciples could be heard wafting through the room's smoky, sad, desperation-filled atmosphere, the only hint of continuity in this world of endless upheaval: ". . . How cold was it? . . . It was so cold that . . . hi-ho . . ." Hoping that Monster had made a jail break, but expecting instead some kind of bad news, Debie peeked through the cracked door. It was Angel, from the unit next door, in her ratty chenille housecoat and big furry bunny slippers with the cartoon ears, crying through her false eyelashes and cradling a baby pit bull. She asked if she could borrow ten bucks. Or maybe Debie wanted to buy her puppy? Debie didn't ask why Angel needed the money—not here at the Motel 6, not anybody, not in this life. "I don't wanna buy him," Debie said. "But I'll take him off your hands." Angel handed him to Debie. "He hasn't eaten in a few days. He's the runt, he can't get to a tit." "Don't worry about it," Debie said. "Kids are sleeping, I'll be right back." Cradling the puppy, she grabbed a twenty from her hiding place under the kitchen sink between two slats, along with a little packet of aluminum foil. "Here," she said to Angel, who was sniffling at the door, exhaling the fumes of cheap wine. "Thanks," Angel said, fading back toward her unit. "I owe you." "No, you don't," Debie said, closing the door softly. Now, at the medicine cabinet in the bathroom, she found a bottle with a dropper and emptied it. The puppy was making little squeaks. "Okay," Debie said, "everything's gonna be all right." Still cradling it, she warmed up some milk, poured it into the bottle, tested it on her wrist with the dropper, then eased open the dog's little jaw, squeezing a drop into his mouth. Johnny had moved into the next part of the nightly ceremony: ". . . some rope, some soap, and some dope . . ." and then came the echo of his priest, Ed McMahon: ". . . some rope, some soap, and some dope . . ." Debie, and millions of other citizens across the land, settled in for the evening—some under the effect of prescription drugs, some plastered on bourbon, some nearly passed out from beer, dying from heroin that was cut with poison, alone in their beds with or without a partner—sinking into her thin lumpy mattress, finally dozing off with her new addition to the family, as Johnny tucked her, and all the others, in. "Let's name him Corky," Mandi said in the morning, picking up the six-pound puppy. "Okay," Debie said, "you name him, you take care of him." "Okay, Mom," Mandi said. "I promise."

By the time Corky was two months and the pride of the Folsom, California, Motel 6, Monster had been transferred to another penitentiary because of a jailhouse fight. Officially, the word was that he was shanked in an argument over food. Debie believed it was one of those fights between two inmates that the guards staged, pitting the prisoners against each other like cocks in the ring until one could no longer get up, betting on the outcome. That's what she heard on the biker grapevine, but who knows? "I gotta get outta here," she said one night while sharing a joint with her neighbor the treasure hunter. "I gotta get away from everybody and everything." He told her he knew a place where no one would bother her, where she could start over, like a baby. "My brother used to date this girl who went to high school in Twentynine Palms," he said. "That's where you need to go." Twentynine Palms . . . It had a ring to it, sounded romantic; exotic and shady. The old warrior showed her a shoebox filled with small rocks. He told her to pick out some gold. She hefted the little pieces of the planet back and forth between her fists, sneezing at the unsettled dust, and settled on a chunk of no particular note other than it had good vibes. The next day she sold some dope, rented a U-Haul and attached it to the pickup, packed up Corky and the kids, then drove to her mother's and picked up Krisinda. Debie told Rose she'd write with her new address. Rose told her to be careful.

"I will," Debie said. And family and dog headed for the desert.

Along the way, Debie tried to think of the good things that had happened in her life. She had her kids. She had a new puppy—a playful thing that was always a sign of a promising future. And she was proud of herself for having survived two marriages to husbands who beat her up. Maybe everything would be all right, she thought as she rolled down the window and watched half the state and the first thirty years of her life disappear in her rearview mirror, as the altitude lowered ever so slightly and she caught the first whiff of the wide-open space that awaited her arrival. Maybe her family could find some peace in the Mojave, maybe this would be her "right turn at Hastings Cut." As soon as they arrived, she promised herself, she would stay away from speed freaks. And then she would never get married again.

A rotation of conventions and conferences having to do with life in the high Mojave Desert passes through Victorville. For a few days during the Underwood trial, the participants of a blasting conference sponsored by Alpha Explosives gathered at the Best Western. Most of them were gold miners who toiled not for themselves as their fellows had in the previous century but for international conglomerates; some worked on road crews around California. Also staying at the hotel were members of an extreme logging team sponsored by the chainsaw company called Stihl. The guys at the Alpha Explosives conference had converged in Victorville to learn the latest state guidelines on such matters as blowing up the land in the name of finding precious metals, and to take a test at the end of the conference so they could renew their licenses for such endeavors. The extreme logging contestants traveled the country, participating in taped-for-ESPN2 events in which they pitted their skills at sawing through massive pieces of raw timber against members of extreme logging teams from other countries who in the off-season continued their careers in other extreme events such as pulling trucks with their teeth. They had stopped in Victorville, on their way from Fontana to Las Vegas, the last stop on the tour before they returned to their regular jobs at various Stihl outlets around the country. It was Taco Tuesday in the Jolly Roger Room. At sundown, a parade of registered guests waited in line at the buffet table for free tortillas and all the ingredients that turn them into tacos when you roll them up. Debie and Jessielyn joined the line, along with Mandi's friend Lydia Flores, who had taken a day off from work to observe the day's courtroom proceedings. As it was during most of the trial, their talk was of the experience of going through it. "That bastard still can't look at me," Debie said of the defendant. "I felt a chill all around me when Gary showed that bloody print," Lydia said. "My arms got goose bumps. It was Mandi." Jessielyn said that her mother would be coming to court soon. After much soul-searching, Juanita had finally decided that she might be able to endure the trial of the Marine accused of murdering her daughter. But Jessielyn was worried. "If you think I get all emotional," she said, "wait till you see what happens when my mother gets here and sees that bastard." A couple of guys in Alpha Explosives T-shirts overheard the conversation. "Well," said a black guy from Yuma, "sounds like you ladies are having a rough day. What brings you to Victorville?" Debie explained that they were in town for a murder trial. That information generally ended conversations, as people were too thrown to inquire further, not knowing the rules of the situation. Lydia noticed that the guy from Yuma was having difficulty processing the news. "There's a guy on trial for killing her daughter and her sister," she said, gesturing to Debie and Jessielyn. "And my best friend." The guy said whoa. His friend picked up the ball. "What happened?" It was a tall skinny white guy from Marysville, Ohio, who worked for a road crew in Southern California. Debie, Jessielyn, and Lydia had assembled their tacos and were moving to the bar. Plates piled high, the guys followed as Debie laid out a few key details. "Some Marine stabbed them thirty-three times," she said. "And he raped them," Lydia added. The black guy said whoa again. The white guy said jeez. The guys from Stihl had been listening and looking for a way to join the conversation, to speak to the ladies, but this was not it. Everyone was silent for a moment. "Your usual?" the bartender said to Debie. She nodded. He mixed up a Jack-and-Coke. The others ordered a beer. "I been waitin' six years for this fucking trial," Debie continued. The Alpha Explosives guys shook their heads. One of the guys in a Stihl tank top, the big guy with the crew cut who was built like a John Deere, nodded knowingly and said, "That's our criminal justice system." The other guy from Stihl, the small, slight one whom it was hard to picture felling giant trees, wondered what caused all the delays. "Oh," Debie said, "there was this lawyer who paper-fucked the system." "Well," he said, "it's our right. It's our right to paper-fuck the system." Debie drank. A waitress stopped by to check on Debie and Jessielyn. "How you all doin' today?" she said. "How'd court go?" "Oh," Debie said, "it went." Jessielyn said that she had a stomachache. The house phone rang. "Bet it's for you," the bartender said to Debie, picking it up. It was. It was Debie's boyfriend Mike, calling from Chicago. He generally called every day after the trial, during happy hour at the Jolly Roger, figuring that was where he could catch Debie. "Yeah, I'm hangin'," Debie said. "Got one of my migraines but I'm hangin'." The bartender pointed to her empty glass. She nodded. He brought her another jack-and-Coke. "Six years is a long time," the white guy from Alpha Explosives said to Lydia and Jessielyn. "Sure as fuck is," Lydia said. "I don't think I could wait six years for anything," the guy said. "That's 'cause you blow things up," Debie said, rejoining the conversation. "You like action." The guys said that they did but that they were worried about their jobs. "It's hard to get permits these days," the black guy from Alpha said. "Gold is used in a whole shitload of things. People don't know nothin' about it. They wanna close up the mines. You ladies wanna dance?" Chuck the doo-wop singer who worked the high desert motel circuit was onstage with a three-man backup band called Shake, Rattle, and Roll. In the 1950s and 1960s, he had sung with the Marcelles and the Orioles. He had a good voice, strong and mellow, and he sang "Under the Boardwalk" and "Blue Moon," and for a few minutes, the gold miners and the members of the extreme logging team and a few Marines who frequented the Jolly Roger after duty at nearby Fort Irwin and some traveling salesmen on their way to Barstow were soothed by the old tunes, even if they didn't know they were old; the Jolly Roger—sometimes rowdy on Taco Tuesday—quieted down and was transported ark-style across the white waves of the Mojave to other places, other times. What was it about those harmonies, those lyrics, that calmed the heart? On her third jack-and-Coke, Debie said that she did not feel like dancing. Jessielyn flashed her wedding ring and the guys from Alpha shrugged and said too bad for them. Lydia had to drive back to Apple Valley—an hour away—to put her kids to bed. When Chuck finished his set, he came over to the bar and hoisted himself atop a stool. "That was nice, Chuck," a waitress said. "Thanks," he said. "Man, I'm weary."

On the first day in Twentynine Palms, Debie found a small bungalow on dusty Manzanita Road, one of the first dwellings with a "For Rent" sign inside the perimeter of Twentynine Palms. It was furnished with well-worn items that, refurbished in the deco stores of Los Angeles, would have fetched a hefty sum, but in the desert were worth only their function, which was nothing when you considered that people scavenged desiccating hulks of cars for the bones of front seats, backseats, anything that would pass as furniture, just like all of the other indigenous opportunists of the bird and animal kingdom who feathered their Mojave nests with flotsam and jetsam. At $200 a month, Debie figured she and the kids could live here for a month and a half before they were out on the streets. But by then, she'd have a job. "Hey, Kristy," Debie said, pointing to the space that surrounded them, "what do you think of our new backyard?" Kristy meandered over some rocks in a wash just behind the house. Mandi looked to Debie. "Go ahead, kids," Debie said, "follow Kristy. I'm not going anywhere." From inside the house, through a dusty shadeless window in the kitchen, Debie watched her three children walk into the desert, toward some scrub, toward their new future. The sun was high in the sky, getting ready to set. It was late August, oven hot, but the quality of the light—its sharpness, the edge of its flight—said that very soon the evenings would be chilly, across the land people would gather the fall harvest in whatever form they took in their yield and begin again. Kristy, Mandi, and Jason would be starting school in a few days.

Rosalie Ortega was born and raised in the Philippine province of Batangas. Tourists often trekked through this area on their way to Mount Taal, an active volcano that produced rich soil for those who for generations had lived in its shadow. But the dollars that visitors spent to have a picturesque view of a natural wonder that was smoldering and dangerous never reached the citizens of this region, and for all the fertility of the land, Batanguenos for the most part were a poor lot whose only way out was the American military. Americans had shed rivers of blood at Corregidor and Quezon and Luzon—all in Batangas—and had maintained a major base in Manila since World War II. Since then, soldiers had been helping themselves to the stock of Filipino women, some of whom wanted nothing more than a plane ticket to the States.

As was customary in her province, Rosalie Ortega's mother Juanita had entered an arranged marriage at thirteen. Nine months later, Juanita had her first child, Ray. Then came

Jessielyn, and then Rosalie. Juanita's husband left and Juanita moved to Manila to make money. The kids moved in with her parents. Even by local standards, getting by was a struggle. Five people now lived in a one-room jungle shack made from logs—batang—left by timber crews. They slept on bamboo mats. There was no electricity and meals were cooked outside on rocks. There was no running water; the Calumpang River a few hundred yards away supplied drinking water, fish for breakfast, lunch, and dinner, and it was also where the family bathed and washed clothes. Some people in Batangas were considered well-to-do; they had generators which provided not only artificial light but power for television sets. Sometimes Jessielyn and her younger sister Rosalie would pay ten centavos to stand outside a neighbor's window and watch *Happy Days*. They fantasized about living in the States, although they had no idea what it looked like there. They wanted to dress like Chachi. In Manila, Juanita was going to seamstress school, and sent back money and handmade articles of clothing when she was able. But there was no disguising the fact that Jessielyn, Rosalie, and Ray were the poorest kids in their neighborhood; at school, the kids made fun of their threadbare garb.

Their grandmother was preparing them for a new life. When Jessielyn was twelve, a young boy began courting her. Every night he would stand with his guitar outside her window under the overgrown banyan trees and the vast, rotting web of vines and deliver a serenade. "Jessielyn," he would sing as she lay on a mat with her younger sister, trying to move the air with a little cardboard fan, "my pretty Jessielyn . . . The stars are bright . . . I wish that tonight . . ." For generations, maybe since time began, this was the prelude to a marriage. Jessielyn's grandmother, fifty-five years old, one of nine brothers and sisters, mother of eleven children, would ask her husband to send the suitor away. She did not want Jessielyn to follow the family example of marrying young. So every night, Jessielyn's grandfather would get up from his mat and step into his one and only pair of denim jeans, come out into the tropical vapors, and discourage the boy, telling him he was not welcome. The boy would warn that soon Jessielyn would become a spinster. "Don't you worry," her grandfather would say. "Jessielyn's going to America."

Rosalie and Jessielyn would giggle and plot their adventures in the land of their dreams. At night, as the girls read fashion magazines by flashlight, Rosalie would fancy herself skinny and glamorous, like the models in their old, well-thumbed copies of *Vogue*. Sometimes Rosalie would cut out the pictures and store them in a special tin, along with other snapshots and clippings that whispered of the good life across the Pacific; to the dreamy girl in the Filipino outback, the pieces of fame and celebrity were stardust, and she knew that someday, she and her sister and brother would experience their own happy days.

After a couple of years in Manila, Juanita married an American serviceman and moved to the United States. A black Marine, her husband was assigned as a drill instructor at boot camp on Parris Island, South Carolina. Juanita got a job making wedding gowns at a little store near the base. Several months later, she started sending for her kids, one by one, because she could not afford airfare for the three together. Eleven-year-old Rosalie, the baby of the family, was the first to leave Batangas. Then, since the girls were so close, Jessielyn followed. When her day came, her grandmother helped her put on her special blue gingham jumper that Juanita had made and sent from South Carolina just for Jessielyn's trip, helped her buckle up her patent-leather Mary Jane shoes that some of the local kids saved up and bought for their friend so she could look just like an American girl. Her grandfather pinned a tag to her dress that said MY NAME IS JESSIELYN ORTEGA AND MY MOTHER JUANITA IS AT

PARRIS ISLAND in case she ended up in the lost and found at an airline terminal—she didn't speak English. Then her grandparents kissed her good-bye and she climbed into the pickup truck that belonged to the neighbors with the television set that ran off the generator. "Bye!" she called as she was driven off to Manila. *"Paalam. Mahal ko kayo."*

The first leg of her trip, a fifteen-hour flight, took Jessielyn to San Francisco. Her connecting flight to Atlanta, the second leg, was late, and the dazed little girl wandered the airport, hungry, scared, and not knowing how to talk to anyone at the dozens of concession stands which offered treats to those who spoke English. A young Marine on his way to Okinawa spotted Jessielyn as she stood before the Burger King counter. He read her tag, realized that she was part of a Marine family, took her hand, and sat her down in a booth under the harsh white light of the burger emporium. He left for a few minutes and returned with a Whopper, urging Jessielyn to unfurl the wrapping. She stared warily at the peculiar packaged item. It smelled funny and she crinkled her nose. "Look," he said. "This is how you do it." He began to remove the paper, and as soon as Jessielyn caught another whiff, she threw up. Then she started crying. He cleaned up the mess and tried to calm the foreign traveler, not knowing that she was used to fish and rice and that the smell of dead meat—her first meal in her new home—made her sick. "You'll be okay," he told her as he waited with her at the gate, perhaps thinking of his own new life in Okinawa, perhaps calming himself. "Everything will be fine."

When Jessielyn finally arrived in South Carolina, she had been traveling for almost forty-eight hours. The deeper she penetrated the States, the more frequent came the bold announcements of American life. First plane ride, first fast food; now a big, long, green car pulled up to the curb where she waited with her bright plastic bags of clothing. The passenger door swung open and there was Juanita behind the wheel. Jessielyn threw herself across the front seat and into her mother's arms. She wondered if the family was living in the car. In Tagalog, then in English, her mother explained that they were not home yet, but the long trip would soon be over. The drive in the roomy vehicle, the cool comfort of the air-conditioning, the shifting of the gears, made Jessielyn dizzy. Juanita stopped the car and rolled down the windows, pointing to a grove of mangroves near the highway. "See?" she said. "It's not that different from Batangas." But of course things were very different; there were even more surprises at her new home on the base: the family had its own television, a refrigerator, a washer and dryer; no more going down to the river to catch food and bathe. But best of all, Rosalie, Jessielyn, and Ray each had their own bedroom. Base housing would not impress middle-class civilians, but to these emigrants from the land where the average yearly salary was $600, it promised even more than they had dreamed of during those damp and steamy nights of watching American TV through a neighbor's window.

As Juanita would tell it, her husband began beating her shortly after her children arrived. But she took the beatings, until she could bear them no longer; she wanted her children to have access to base medical care. In 1985, soon after the family had been transferred to Hawaii, the marriage fell apart. Ray had just graduated from high school. Jessielyn was now a junior and, having avoided her native tradition of marrying at puberty, was about to marry a Marine before she finished twelfth grade. Rosalie had dropped out of school as a sophomore, moving to Southern California and hooking up with a black Army serviceman called Burnelle. She sent a letter back to Jessielyn in Honolulu. She was pregnant, she wrote, and her baby was due in one month. And she also mentioned her new tattoo—a Playboy bunny, on her ankle. "Maybe you can get one, too," she wrote.

Soon Juanita met the love of her life, a Marine named Tom

Brown. He was fifteen years younger, and white. He took her from Hawaii to his home in Michigan. She was not acclimated to the cold weather, and even though she could go anywhere she wanted to in the family's ancient four-wheel-drive, she felt trapped. After three years in the upper peninsula, where ice fishing was a way of life, she wanted to return to her warm hut in the jungle of Batangas. She and Tom decided to move to Southern California to be near Rosalie. Now out of the Corps, Tom had lined up a job as a long-distance truck driver. Juanita, still working as a seamstress, wanted to join her man on the road, figuring that two truck drivers could make a lot more money than one. So from Michigan to California, Tom taught her to drive a big rig. Juanita was a quick study, easily shifting up and down through the sixteen gears, fearlessly accelerating to pass slower trucks on a grade, soon becoming a road novelty and, at a petite five-three, earning the CB moniker "Asian Doll."

The pair settled in Garden Grove's little Saigon section outside of Los Angeles, where Rosalie was living with her new boyfriend, a white metalhead and ex-Marine named Chuck. She was going to teller school at the Sanwa Bank and had a little girl, Shanelle, who was three years old. Ray soon followed from Hawaii. Jessielyn was married and pregnant. By the standards of Batangas, the girls had violated custom by not starting their families at puberty. By their own standards, they were now free. Their escape from the Philippines had been exactly what they wanted, a flight from poverty and a tradition that kept girls barefoot and in their place.

But their route of entry into American life had placed them squarely on the road that leads to Twentynine Palms. By 1988, they were heading east on Highway 62, very far from the steamy green confines of a poor island in the South Pacific, deep into wide-open Mojave terrain, welcomed by space and promise, senses heightened, a bit on guard perhaps; the Joshua trees looked weird, nothing about the place was inviting. Jessielyn, the first Ortega to stake out desert territory, knew that the jungle was far behind. But what awaited the family? she wondered as she drove past miles and miles of nothing. She had never imagined such a place; it was another, strange American "first."

She came with her husband and her son; the Marines had transferred the family from Long Beach, California, to the base at Twentynine, and they moved into a nondescript tract of base housing on Two Mile Road, a few minutes north of town and south of the base. Juanita and Tom followed. Rosie arrived a couple of months later, without her boyfriend. For a time, she and Shanelle lived with Jessielyn. But Rosie was eighteen years old and wanted her own apartment, wanted to grab on and climb up, on her own. She could tell from the ad in *The Desert Trail*—"cheap" and "close to town"—that the apartment complex at 6422 Palo Verde Road would be her new address. She called Juanita and asked her to meet her there, then borrowed her brother-in-law's Honda Accord and headed right over. Here, a couple of blocks north of Andreas, a Greek diner on Highway 62 favored by Marines, slightly south of the Bowladium on San Gorgonio Drive, and just east of The Alamo laundromat, were eight separate units in an L shape around a driveway, two per building, each with its own outdoor entrance. Just like the ad said, Rosie thought as she parked in front, a good deal, not *Happy Days,* but that would come. A friend dropped Juanita off a few minutes later. "Hey, Ma," Rosie called. "What d'ya think?" At first look, Juanita felt a knot in her gut, something about the place felt evil. "I don't know," she said. "I was hoping maybe you could find a place near Jessielyn." "Come on," Rosie said, heading for unit number seven. "It beats the PI."

Indeed, the one-story stucco complex was a picture of the past, present, and future of American life on the run, of the fate of those who are slaves to starting over. The haphazard stone driveway was lined, almost by accident, or as if attempted in another geologic age, with starving oleander bushes that bloomed in spite of themselves. There was an occasional eruption of ocotillo. Scruffy dogs ranged the turf. A fossilized Pontiac was, would always be, up on blocks—what's a desert habitat without this artifact? Outside various entrances to the bungalows were empty six- and twelve-packs of Budweiser—another sad Anheuser-Busch colony on the edge of nowhere. ZZ Top, the voice of the Mojave, belted an invitation to "slip inside my sleeping bag" from a boom box in one of the diminutive and wretched pads.

A couple of bikers on Harleys pulled up. A big roly-poly guy without teeth opened the door of the apartment where a cab with two flat tires was parked. The bikers went in, bought some speed, came out, left, gunning their pipes. A few minutes later, a girl, about twelve, emerged—rangy, wearing baggies and a T-shirt advertising an old Kiss world tour. She hopped on a skateboard and headed into town. The door marked "8" opened; another empty six-pack hit the pile. The door closed. "Rosie," Juanita said as Rosie turned the knob on the apartment with the sign that said, VACANT—COME IN, "I just saw a terrible thing." Rosie told her mother that she was planning to live here for just a little while, until she was a millionaire, and then they could all live in a mansion in Beverly Hills.

Rosalie Ortega moved into number seven at 6422 Palo Verde Drive a few days later. She was starting off the new year of 1991 just right. The one-bedroom was perfect for her daughter. It came with all appliances, except for a washer and dryer, which Rosie would install later, next to the water main outside her front door. It was conveniently located, just a few blocks away from the Texaco gas station/mini-mart where Rosie had gotten a job as night manager. She would save some money, finish high school in Twentynine Palms, get her American citizenship, and then . . . who knows? Extremely outgoing—"too outgoing," her sister would always tell her—and excited about her new life, Rosie made friends fast.

At school, she started hanging out with some of the girls in Mandi's crew, including Lydia Flores and her two sisters. At the Texaco, she struck up conversations with just about anyone who bought fuel. Since she worked nights, this typically included Marines and gangbangers. "I don't have no money for gas," a Crip in a blue bandanna would say after the bars had closed, calling out at the pump from his Mercury with the smoked-out windows. "Can you hook me up?" "No problem," Rosie would always say. "Give it back to me when you have it." And he would.

When her best friend from Hawaii arrived, broke and lonely, she hired Arlisa at the gas station and let her share her apartment for a few bucks. Her friend was getting a good deal; the sad little stucco cave was, thanks to Rosie, now a cozy and inviting hangout. She replaced the sunburned yellowing window shades with handmade curtains. She covered the Formica table in the dining area with a pink cloth. Evidence that she indulged her daughter was everywhere; amid the clutter of fashion magazines and schoolbooks and nail polish and makeup and compact discs were teddy bears and windup turtles and items large and small bearing the image of Mickey Mouse; like a lot of American kids, at five Shanelle had a pretty impressive piece of the action. But as far as Rosie was concerned, the apartment was empty unless people were dropping by. She gave her address and phone number to just about anyone she met and many took her up on the invitation. There was a steady stream of visitors whenever Rosie was in and awake; strangers and friends alike made themselves at home at 6422 Palo Verde, number seven.

For the lost souls of this town—young Marines far away from their parents, young toughs without parents, kids with down-and-out parents—this was no small gift. If she had the time, Rosie cooked for them, making lumpia, kind of a Filipino egg roll, or stir-fry. She had all the best rap CDs, and she would crank up the box and there would be dancing until the last person fell asleep or went elsewhere. She was known for giving her girlfriends elaborate manicures, making them forget their troubles in the pretty acrylic pictures of palm trees and beaches painstakingly applied to their long nails, comforting the girls who took care of their brothers and sisters, took care of other kids in the neighborhood, mothered their parents (usually their mothers; the fathers were generally not on the premises), mothered their own children, took care of the children of their friends. Very quickly, there was a wide range of locals hanging out at Rosie's.

Often, her mother would stop by and pick up Shanelle, taking the young girl on her truck runs with Tom. "I don't want Shanelle around all those Marines," Juanita always told Rosie. A lot of the Marines were black, but Marines of any color would have concerned Juanita. Sometimes they carried large stuffed animals that they gave to Shanelle. Sometimes they brought groceries from The Alamo for Rosie. She introduced them to her friends, including Mandi Scott, who often baby-sat for Shanelle. It was one big, extended family. "You know that this place gives me a funny feeling," she said. Rosie knew that by "this place," she was referring to the entire apartment complex, filled for the most part with the unsavory characters she and Juanita had seen months ago. Juanita was not alone in her suspicion. The apartment complex was so popular with certain Marines, especially members of the 3/11, an artillery regiment that had served in the Gulf War, that it was under the surveillance of the Naval Criminal Investigative Service, or NCIS as it was known, the cops who investigated Marine crime and occasionally tailed Marines when they went into town. Sometimes two nondescript guys with leatherneck haircuts sat in an unmarked government vehicle across from the tiny community and watched people coming and going. One evening, as a Michael Jackson tape blasted from inside her apartment and Mandi and a few Marines tried and tried to teach Shanelle how to moonwalk, Rosie came out with a platter of hot food and approached the G-men in the Ford. "Want some lumpia?" she said. "No thanks," they said. "Never know what's really in there." "Yeah," Rosie said. "Better watch out for home cooking."

In May of 1991, Rosie threw herself into preparations for her July Fourth bash. Even though it was early, she was excited; this was her first party in her own apartment, and it was a very big deal, especially in a military town where everyone really knows how to ring in the birth of the nation. Although she never wanted to go back to the PI, felt completely at home in Twentynine Palms, she wanted to serve up a feast that would make her grandmother proud. Rosie spent the next few weeks turning her apartment into a "kainan kamayan," a place where you eat with your hands. She made vast quantities of lumpia, which she stored in the freezer. Knowing that her American friends would bring burgers to throw on the grill, she cooked a pot of bagoong (fermented shrimp paste for barbecuing meat). She went shopping on base and stocked up on jasmine rice, Dr Pepper, Betty Crocker cake mixes, red, white, and blue cake decorations. At Stater Brothers, she bought cases of Olde English beer. She told her friends to bring their favorite rap tapes, on the off chance that she didn't have them herself.

But later that month, as spring fever reached its pitch in Twentynine Palms, and baby rattlesnakes and tortoise hatchlings and young coyotes and all creatures great and small were emerging from their burrows and nests and skittering and clambering around and outside of town, another presence slithered out of the depths. Some people say that the devil lives out here. After all, it was the desert, wasn't it, where ancient wise men wandered in order to confront darkness, to go one-on-one with the putrefying guy with the pitchfork and horns. Other people, people who want to distance themselves from bad news, from people who are down on their luck, call the presence trash. No matter the name, a murder occurred in Rosie's universe of all-night gas stations, fast food, and the working poor of all races and backgrounds. That it occurred in this world should not have been surprising because those who live in it have lives that are rife with physical violence; as often happens with marginal people who live moment to moment, someone's cousin was always in jail, someone's friend just had his teeth knocked out in a bar fight, someone's sister just got punched in the gallbladder by someone else's adulterous husband. But such is the nature of murder that even those who reside near it are shocked when it happens. In this instance, Arlisa was working the night shift at the Texaco. A Filipino girl—and there were many in town; like Rosie and her sister and mother, they had married Marines—ran across the driveway. A black guy was chasing her, grabbed her, started beating her up. Arlisa called the cops. There was a Mexican guy at the pump. He tried to stop the fight. The black guy smashed the Mexican's head in with a rock. A few days later he died. Arlisa's testimony put the black guy away on a charge of manslaughter.

After the incident, Rosie told Arlisa to move out. Asking a friend to leave was uncharacteristic of her, but Rosie did not want to live with someone who had seen a murder unfold. It was bad luck, it would take weeks for the curse to leave her apartment. Maybe she should cancel the party; it would probably be a disaster. Arlisa called a Marine and asked him to come pick her up. "Hi, Val," Arlisa said as the big Marine pulled up in his Cadillac. "Let's go," he said, helping her with her bags and winking at Rosie as he escorted Arlisa through the door. "Hey, Val-en-teen," Rosie said. "What's up?"

Like many who found themselves in the Mojave, Valentine Underwood's path to the desert was written long ago. The youngest in a family of seven children, he fell far from the tree. He and his brothers and sisters were raised by strict parents in a Baptist household where Bible reading was a daily activity. One of his favorite Bible passages was the story of Daniel in the den of lions, which his mother would often read and talk about as a parable of racism in America. "Remember how Daniel was falsely accused," she would say. King Darius sealed Daniel away in a cave with lions to test his faith. Then he left for dinner. The following morning, the king rushed to the sealed door. "Daniel," he called, "servant of the living God, has your God, whom you serve continually, been able to rescue you from the lions?" Daniel replied that God had sent an angel, who shut the mouths of the lions. Darius unsealed the cave and hailed the servant's faith. The men who had falsely accused Daniel, along with their wives and children, were thrown to the lions. "Right will always triumph," Zimena told Valentine. "God looks after his children wherever life may take them." Years later, some of Zimena's children suspected that the Bible passage had a personal meaning for their mother, something that had as much to do with the state of her marriage as with social conditions. She was a hardworking nurse, the parent the kids counted on. Her husband David was a jack-of-all-trades who would sometimes disappear at night. Zimena speculated that he had a secret life. She and David divorced after nineteen years, when Valentine was seven. She and the children moved to a lower-middle-class neighborhood in the Baltimore suburb of Weaton. After high school, five of the Underwood kids entered branches of the military, including the Marines, and flourished there.

An average student at JFK High, Valentine was a promising basketball player and hoped to join the NBA. At six-six and 240 pounds, he made a powerful center, and had the moves to back up his size. While growing up, he modeled himself after Patrick Ewing, then star of the Georgetown Hoyas; later, he took his idol's number—33—as his own. He enrolled in 1981 at Lubbock Christian University on a basketball scholarship, figuring that it wouldn't be long before he was scouted and recruited for professional basketball.

But his record became notable mainly for his mistreatment of women. A number of them lodged complaints with school authorities, reporting that they had been sexually or physically assaulted by Underwood. Others reported harassment. In 1985, because of his behavioral problems, he was asked not to attend LCU for one semester. He used the time to take some ROTC courses at Texas Tech. In 1986, he returned to complete college, graduating as a psychology major with a 1.95 average and a reputation among school authorities for being a manipulative student who intimidated and controlled peers and faculty members.

Disappointed that he had not been recruited by the NBA, Underwood left the Bible Belt and headed back to the East Coast to find a job. For a while he worked at a Jewish home for the retarded in Takoma Park, Maryland. On July 2, 1987, he was arrested by the Montgomery County Police Department for rape. Shortly after the incident, the victim disappeared and the case was dropped. Underwood then passed some time in upstate New York. On June 6, 1988, in Binghamton, he was again arrested for rape. Once again, the case was dismissed for lack of evidence. He returned to Maryland, where he worked as a civil inspection engineer until he was fired in 1989.

After failing elsewhere, there were only a few places left for a guy with his résumé: Burger King, the street, or the military. He chose the Marines. It was a match, for in subtle ways, they chose him. On the face of it, he was exactly the kind of guy for whom the Corps trolled.

At the time Underwood enlisted, his age of twenty-nine was not a mark against him, but just the way things were: the ranks of the Marines were filling with older recruits, many of whom had been laid off from factory jobs. The Army, Navy, and Air Force promised money, technical training, travel, and tuition loans to attract recruits. So did the Marines, but these enticements were not the heart and soul of the Corps' advertising campaign. The Marine Corps offered a challenge, promising only a hard time. The subtext for the rigors of Marine boot camp was: Can you stand up to the legacy of the thousands of Marines who died in action at Inchon, Belleau Woods, Iwo Jima, Okinawa, Guadalcanal? Can you spill your blood if called? Do you want to be the first to fight and the last to leave the battlefield? You've never shied away from a physical confrontation, have you? Wouldn't you describe yourself as a warrior? Isn't it time the country had some heroes? The dare for "the few, the proud" to come forward has been a successful one: for the past fifteen years, since the end of the draft, the only branch of the military to meet its monthly recruiting goals has been the Marine Corps.

Valentine Underwood, an athlete who thrived on rising to a challenge as a good athlete must, a guy who thrived as part of a team and identified with a number popularized by an older role model, a guy who wanted to be drafted by the NBA—an entity which for all of its athletic talent had become nothing more than a hollow image of doing something cool (completing a slam dunk) while wearing something cool (a pair of Nikes)—took the Corps up on its challenge. On sizing Underwood up the day he walked into a Baltimore recruitment office, the recruiter undoubtedly caught the drift of this wandering soul. But he may also have been charmed by Underwood's easy smile and soft-spoken manner. The big guy did have a way of ingratiating himself. Or maybe it was something more basic; maybe it was time for the recruiter to make his next house payment and he needed a bonus so much that when he looked it up, he ignored Underwood's history of arrests for rape. Or maybe lots of recruits had such histories, and as long as there were no convictions, it didn't matter. In any case, circumstances had converged in such a way as to permit Valentine Underwood to join the Marines on waivers.

Certainly the enlisted ranks of the Corps had long attracted those whom the upper classes viewed as unsavory—that was its pride and its shame; in many ways, the Corps invented *The Dirty Dozen*. Before the Revolutionary War, the colony of Virginia openly used its regiment of Marines as a dumping ground for debtors, vagrants, and criminals. In 1775, conscription for the Continental Marines went right to the source: officers accompanied by drummers scoured the bars of Philadelphia, leaving with platoons of drunks. After the Revolutionary War, the enlisted population expanded in number but not class, composed of unemployed laborers, runaways, teenage apprentices, farmers, craftsmen, itinerant schoolteachers, and clerks. By 1859, the ranks filled with more runaways (mostly young kids from farms), more of the unemployed (mostly from cities on the East Coast), Irish and German immigrants, drunks, fugitives, debtors, and career sergeants and corporals in their thirties and forties. In early World War I, enlisted men were mostly city toughs and farm boys. But as Ernest Hemingway wrote passionately of spilling blood for freedom, the literary elite rallied to concepts of "honor" and "duty." The demographics of the Corps changed dramatically in World War II, with academics, intellectuals, doctors, and lawyers joining up to fight the enemy. Hollywood saluted the leathernecks in movies like *Halls of Montezuma*, *Marine Raiders*, and *Pride of the Marines*. But after the Allied victory in 1945, the Marine rank and file began a return to its scrappy origins. By the time the war in Korea was over, it had lost what little remained of its appeal to the country's intellectual and professional establishment. In the 1960s, as the country began its war against North Vietnam, the Corps once again swelled with recruits from the poor and working class—in the latter part of the twentieth century the only members of society to heed the call to arms, the only members of society to respond favorably to calls for sacrifice, the only members of society who did not seek and were not granted deferments.

On June 6, 1990, under the iconic photograph of six Marines planting the American flag at Iwo Jima, a recruiting officer shook hands with Valentine Underwood and bid him good luck in the Corps.

Sooner or later, on their circuit of base assignments, most Marines pass through Twentynine Palms. For Underwood, it was very soon: things were about to blow in the Middle East and in preparation for the run-up to the Gulf War, Desert Shield, troops were sent to Two-Nine to train for desert warfare. The vast acreage of the base stretches from Twentynine Palms north across the Bullion Mountains to the empty town of Ludlow, just south of Interstate 40, and west from the town of Landers on Old Woman Springs Road to the beer, gas, and lodging pit stop of Amboy, just east of the base boundary. Aboard the base, Marines in the field lived at Camp Wilson, a training ground at Deadman Lake. The surrounding rocks bear petroglyphs made by the Serrano Indians thousands of years ago. Only certain shamans were permitted to draw the glyphs, recordings of what they saw while in a trance, which took them to the world between here and there. The Marines who trained in the desert for other desert wars also made a permanent mark of their experience. They engaged in Combined Arms Exercises (CAX), learning to move on the land with accompanying artillery and air support, learning to destroy the desert. To the eye that is disconnected from the soul, the heart whose ache is elsewhere, the Mojave invites destruction, cannot convey its essence: "Take care of me. I am fragile. I bleed. Sometimes I am pretty." It is almost asking to be raped; the sands are pocked with spent shells, grenade caps, ancient ordnance. Virgin guns are broken in on a still desert evening; aging guns are fired one more time before retirement ceremonies. If anyone knows that Kilroy was here, it's the Mojave Desert, and because it received Kilroy so willingly, the CAX that were played out on its haunches were exported to its distant yet very close cousin; the Gulf War transformed the desert of Kuwait into a distant mirror of the Mojave, trashing its own peculiar treasures, setting its oil fields ablaze, crushing its life beneath advancing tanks and artillery trucks, dispossessing it of its beauty.

"Hey, handsome, what can I do you for?" Debie McMaster said to a young Marine at the E Club, where she had been working since shortly after her arrival in Twentynine Palms. She felt at home at the E Club, a hangout on base for enlisted men, it reminded her of many of the men she had known and loved in her life, and she knew how it was around them. "I'll take a Seven-and-Seven," the jarhead from Florida said. As soon as she served it, he downed the whole thing, then asked for another. "Long way from home, huh, junior?" Debie said. "That ain't it," he said. "I can't stand the desert." Debie agreed that the desert was different. The Marine downed the next drink and said, "I never seen a place with so much nothin'." Debie replied that sometimes nothing is a good thing.

For the next two years, nothing happened—nothing that would be marked as an insult to the system of Debie's family. What did happen was all good, and new: Debie flourished, on her own, acquiring a devoted following of bar patrons. The kids found new friends among other distressed families in town, and settled into their schoolwork. As long as they maintained average grades, Debie was satisfied; like a lot of kids in town, they weren't the kind who did well on tests. But it was not easy to shake her past. A couple of years after arriving in Twentynine Palms, some of Debie's old running mates sent word that they needed a favor. They had made an arrangement with a Luiseno Indian known as Russell who lived on the Rincon Reservation outside San Diego. The Rincon Indians were poor, like many Indians, although not as poor as some; in addition to a small income derived primarily from bingo, they earned money from the cultivation of the prickly-pear cactus, a staple of local land development. Russell was a cooker of methamphetamine; some of the reservations in California had become safe havens for such illegal activity as they were difficult for the feds to monitor. Russell, Debie was told, was moving speed off tribal land. Along with some bikers, he had taken over key markets in the area. Now, the plan was to cash in on the insatiable high desert habit. How would Debie like to make a lot of money really fast? There was protection, Debie was told; they assured her that she would not get busted. Debie knew this might or might not be true. In either case, it didn't matter. Certain people had bailed her out when she was on the skids, many times in fact, and she owed them big time. She was taking a risk, she knew, but if ever there was anything the desert was for, taking a risk was it. The minute she had hit town, spotted a few edgy-looking drifters, heard the bark of a neighbor's rottweiler, she knew that here was a place

whose "don't ask, don't tell" policy was right in sync with the way she was raised. So she began driving every week, out of the Mojave and down toward Mexico, to the reservation where the view from its famous Mount Palomar told the violent and beautiful tale of California: looking west, you could see all the way to Camp Pendleton and the Pacific beyond; east, into another spectacular desert, the Anza Borrego. But Debie never lingered for the scenic vistas at Rincon. She picked up the speed and immediately returned to Twentynine Palms. It was quickly sold. But of course, the enterprise was destined for failure. A teenage speed freak had been living with Debie. She and Debie had met long before at the Motel 6 and one day, after having crashed at Debie's another time, the girl showed up again with her toddler. She and her son needed a home. Debie said yes, of course they could live there. Like other young women who roamed the desert, this one was often in trouble. There was the time she sold half an ounce of speed to an informant and climbed out Debie's window as the cops arrived, spraining her ankle as she fled and calling Debie from a pay phone to ask for a ride home. Or there was the time when things got really rough. One night, a few minutes after Debie had returned from the late shift at the E, the phone rang. It was Russell, with a tip. The cops were on their way. Debie grabbed her scales and bags of speed and drove to a Dumpster behind a nearby grocery store. When she returned there came a loud knock at her door. Corky jumped from Mandi's lap and ran to it, staring at the door, growling. "Police, open up," a man outside said, over the loud voice of Don Cornelius on *Soul Train*. "Mandi, turn that down," Debie said to her daughter, "and take Corky out the back." Then, to the presence outside: "What do you want?" Debie said. "We want in." "Do you have a search warrant?" Corky's growls became more threatening. Mandi tried to quiet him. "I said do you have a search warrant?" Debie repeated. Corky wouldn't budge. The

door blasted open with a kick to the lock, and two cops barged in with guns drawn; the pit bull lunged—"Corky!" Mandi yelled—a cop feinted a move to the dog's left, and the other one held a gun to Mandi's head. "Get him out of here," the cop said to Debie, and Debie said, "He won't move until you take that gun away from my daughter's head. And if you kill him, I'll get you for animal abuse and violation of my Fourth Amendment rights. Unless you got a search warrant." "We don't need one," the cop replied, and said they were looking for Debie's houseguest. "Where is she?" Debie said she didn't know. "Who's in the shower?" he said. Debie shrugged. One cop headed for the bathroom while the other continued to hold the gun at Mandi's head and Corky continued his vigil, fangs bared. The girl was in the shower, smoking speed. The cop dragged her into the living room. Then he tore the place apart. He found nothing, other than the small quantity of speed that the girl had almost finished. "What about my rights?" Debie said, reiterating the time-honored cry of the desert, of those who made this endless PO Box their home: "WHAT ABOUT MY RIGHTS?" "You gave those up when you hooked up with the wrong team," one of the cops said as they left with the girl. Mandi told Debie that they didn't need a reason for the cops to return. She grabbed Debie's bong, the one with fairies dancing on top of a mushroom cap, and ran outside and smashed it against a rock. "Hey," Debie said, "that's my favorite bong," but she knew that Mandi was right: she called the reservation and said she was out of the business. A few days later, there was a drug sweep across the high desert. It did not involve Debie. She thanked her zodiac sign profusely and hastily threw together a little shrine of candles before some tiger postcards she tacked up above a kitchen counter. She had already used up about a thousand of her nine wildcat lives and wondered how much longer she could stay one step ahead of the forces that shadowed her family.

"Guess what happened?" Mandi said to her friends. "Guess what happened at my house the other day?" The story of the raid and how Corky had come to Mandi's defense became a key part of Mandi's repertoire, embellished over time and employed to impress friends in a circle of people where defiance was the coin of the realm. But the near miss reminded Debie that she wanted more for her second daughter. In fact, all of Debie's hopes for the family were riding on Mandi; now, after seven years in Twentynine Palms, Mandi was fourteen, growing into a young woman who resembled her mother in almost every way. Mandi was also a Leo; mother and daughter shared the same intense green wildcat eyes that they proudly identified with their astrological symbol. In high school, Debie had been an average student. So was Mandi, like her mother not responding to the constraints of traditional education. (Although Twentynine Palms has its share of boosters and proud parents of students who always get good grades, amid all the Semper Fi decals on cars it's not uncommon to see bumper stickers that say, MY KID BEAT THE SHIT OUT OF YOUR STUDENT-OF-THE-MONTH, sometimes on the same vehicle.) But mother and daughter excelled in the physical world: Debie was a tomboy; Mandi, too. One of Mandi's prized possessions was a Steve Garvey poster she kept on her bedroom wall next to her bat and glove—like Mandi's idol, she played first base. She was also a member of a local swim team called the Aquatics, and proudly displayed her ribbons (Mandi did not know that her mother had been on her high-school swim team until after she started competing). Both Debie and Mandi loved to dance, to party, though Debie's preferred soundtrack was white rock and country; Mandi's was rap. Both Debie and Mandi loved kids. Debie occasionally took runaways in off the streets, along with stray dogs, though the animals were not generally welcomed by Corky, and they did not stay long. (Once, while Corky was hidden at a friend's for several days, Debie had gone to jail in his place, to prevent his arrest by animal control authorities who had received complaints about his marauding.) Mandi was a frequently requested baby-sitter around town, in fact known by many as "Aunt Mandi," often taking care of toddlers whose mothers were her age or perhaps a little older. Generally the kids had no fathers, or stepfathers who came and went and had other children to worry about anyway, or the fathers were Marines and stationed abroad. Some of the mothers were methamphetamine freaks, availing themselves of the ample desert supply (almost every week the local papers carried news of an exploding trailer, which always meant a makeshift speed factory had gone up in flames somewhere in the Mojave). The kids always responded to Mandi, who brought her rap tapes and special books like *Pinocchio* or *Charlotte's Web;* for some five- or-six-year-olds, these were the first books anyone brought into their homes. Both Debie and Mandi were fierce allies, terrifying foes, and had themselves engaged in a couple of knock-down-drag-out fights. At five-six and 160 pounds, Mandi had the advantage, but in contests between mother and daughter, she would ultimately back off and obey Debie. By that time, the rage had moved on like a desert dust devil; the two of them always calmed down as quickly as they were possessed.

Sometimes Debie would take Mandi for a drive in the desert. They would go to a favorite place among the Joshuas, climb a secret trail past a patch of lupine and primrose and poppies to the top of some rocks, and gaze across the wide expanse of the Mojave. Mountain girl Debie never thought she would adjust to the desert, but finally she was beginning to feel like this was her home. The desert got under your skin if you stayed here long enough: Debie liked to watch the shooting stars at night and listen to all the critters howl, and by now she had her own Mojave moniker—"Desert Debie"—a sure sign that she was a local. Things were going pretty well for her family; Mandi would be graduating from high school in three years, Kristy was planning to join the Army, and Jason was getting by.

MESS WITH THE BEST, DIE LIKE THE REST! said the bumper sticker on the back of a Chevy pickup. Although it bore the Marine logo of the eagle atop the globe and anchor, this was not a government-issue bumper sticker. It was a popular item in the local variety store, along with "Death Before Dishonor" and "Marines Don't Just Read About History—They Make It." These stickers and tattoos and decals—extreme to outsiders—expressed the famous Marine esprit de corps, a mind-set that set them apart from, say, the Army, whose members often announced themselves with Mickey Mouse tattoos. The truck snaked its way through the back roads of Twentynine Palms, today clogged with traffic, looking for a way to hook up with the massive military advance that was making its way into town from the 10, across Highway 62. The people in the truck, wives and girlfriends of Marines, were all decked out in their most alluring tube tops, tight jeans, high heels. In the back of the Chevy were four teenage girls in stretch denim and T-shirts that said LUNCH BOX GANG. For the first time in eight months, they, and just about everyone else in town, were happy: their boys had just kicked Saddam's ass in the Gulf War and now were coming home to the Mojave Desert, where they had trained, and trained well, for the desert operation. In fact, the military and the desert—well, they had a thing going on. In World War II, General Patton trained his troops in the Mojave in preparation for the desert battles in North Africa. The training was so successful that it gave rise to the military saying "We do deserts, not mountains." In fact, tracks from World War II tank maneuvers are still visible in the sands to the south of Twentynine Palms. Returning to the Mojave was not just a homecoming for those who fought in Desert Storm, but a communion with one of the proudest moments in American military history. And then of course, there were the two things that were of most concern to all who longed for this moment—sex or money.

For Mandi Scott, now fifteen, it was money. Debie had been tending bar at the Iron Gate, one of sixteen bars in town, and ever since the boys went to Saudi, the tips had dwindled down to a few bucks per week, courtesy of whatever the cash-starved locals could extract from their pockets. More than once, Debie had been tipped with food stamps. But she never used them because she was too embarrassed. Not that the family didn't need them. Pretty much since the deployment of troops to the Middle East, Twentynine Palms had been a ghost town. Most of the stores were closed or had scaled back hours. Homes, furniture, entertainment units all over town were being repo'd every day; even life on the installment plan wasn't cutting it in the low-cost Mojave. Tourism took a hit, too; the hotels were empty because there had been no wildflower blooms that season. The only signs of life were colors and sounds: the neon of bar signs—THE JOSH LOUNGE and THE VIRGINIAN and DEL REY'S—which looked so pretty against the dusky desert skies, betraying no hint of the desperation unfolding inside, and the occasional rumble and then sweet fade-out of Harley pipes on Highway 62, hightailing it to Yucca Valley, where the gas station was open twenty-four hours—and wasn't this why we were fighting the Gulf War in the first place? But here in the town that was home to the world's biggest Marine base, residents were stuck. Mandi, Krisinda, and Jason weren't the only kids in town eating macaroni and cheese for dinner every night. Having to serve it up all the time humiliated Debie and all of the other working poor in town; the local edge felt sharper, people sucked harder on the last of their cigarettes, as if the inhale would take control of things, and then they downed pitchers of beer, because they knew it wouldn't.

The driver of the pickup approached the convoy of tanks, Jeeps, artillery guns, buses and started honking furiously; there they were, all the scruffy-looking guys, when they left they were kids, now they were war heroes (though having engaged in no combat) who had saved America, the world, and they acted in the manner of all returning victors: as the crowd waved flags, and red, white, and blue streamers, and placards that said, HEY SADDAM, EAT MY DUST, as some spectators ran to the buses crammed with war vets and proffered six-packs, the boys in their cammies reached down through the bus windows to slap five, grab beers, grab some tit—for eight months they had lived in a world where women were "off-limits," covered with veils, and some now contorted through the windows backward, mooning the welcoming crowd.

The desert two-lane was festooned with eighty miles of yellow ribbon—all the way from the interstate to downtown Twentynine Palms where 62 became Main Street and intersected Adobe Road, the major north-south thoroughfare that led to the main entrance at the base. The red truck carrying Mandi and her friends from the Lunch Box Gang fell in line at the end of the parade and followed the boys for a while. They looked good—all grungy, tanned, roughed up, hardened, if only from living in a desert bunker for thirty-two weeks, eating MREs, and traffic-copping Iraqi POWs down the line toward the detention area at the back of the advance.

"Hey, boots," Mandi shouted, "you rule!" A couple of the guys responded with a nasty, "Oorah."

"Mmm, mmm, mmm, look at that fine desert scenery." This was Lydia Flores, one of Mandi's best friends, beautiful, sixteen, tough talking, and self-possessed in spite of her hardly menacing "Lunch Box Gang" T-shirt.

"Not for me," Mandi said. In other words, not black. Mandi, who was still carrying a few pounds of baby fat, was into black guys, and they were into her. Black guys liked the extra poundage, and also the fact that she liked to dance.

"You seeing Kevin later?" Lydia asked.

"Yeah, he's coming to the party at the Gate. If my mom doesn't throw him out." Debie was not happy about Mandi's choice of color in boyfriends, and in fact often told her daughter not to "burn coal." "Why do you care so much about color?" Mandi would say. "What difference does it make?" But in the high desert, Debie was not alone in her view; for all of Mandi's refusal to choose friends based on skin color, there were plenty of others who did just that. In the Mojave, for instance, there was the unspoken tradition of stopping black people on the freeways for being a mile or two over the speed limit, while never stopping white people, unless they were driving spent muscle cars held together with duct tape, which generally translated as drug dealer. There was a well-known situation in certain restaurants, in which minority patrons who desired a table would be kept waiting for hours while white people were immediately seated. And even Mandi's own brother Jason, attempting to stake his own identity, fell in with a white-power crowd at school, after being repeatedly taunted by black kids and "wiggers" (white kids who fancied themselves gangbangers from Watts) as a "Hessian" because of his buzz cut. Indeed, as America and the world retreated into genetically coded tribes, the Lunch Box Gang was a breathtaking manifestation, a cross-cultural "point of light" that neither President Bush nor his speechwriters would ever behold.

Mandi's friend Tina, also in her LBG T-shirt, suggested hopping off the truck and racing to the base entrance so they could personally greet the guys when they got off the buses. Her reason for being happy that the boys were back had nothing to do with sex or money. A good friend had been in the Gulf and she was happy that he'd come back alive. She thought maybe she could give him a hug if she was lucky enough to be at the right bus at the right time; he didn't know that his father was in jail on another DUI and his mother had cracked up three months ago. The girls took off, easily outpacing the parade convoy, high-fiving friends in the crowd, four teenagers skipping for joy, adding to, driving the swarm of jubilation, Mandi especially hating it when her mother felt low, loving it that her mother's tip jar would once again overflow, for she was the most popular bartender in town, that is, when anyone was in town. Now, like the desert frogs that manifest after a rainstorm, Twentynine Palms would once again come alive. At the base gate, the Lunch Box Gang, Mandi leading the way, squeezed through the crowd past weeping women, ecstatic women, seductive women, until they got to the front of a bus, only to find that another member of the town's greeting party had planted herself at the doorstep, hoping to get kissy-face with a Marine she liked. This was twenty-year-old Rosalie Ortega. Rosie liked black guys, too. "Hey, homes," Mandi said, "waitin' for the man?" "Yeah," Rosie said, hugging each member of the LBG, "I paid my cousin to fill in for me at work. I'm losing money on the deal, but hey"—and now gesturing at the Marines—"check out all these rockin' bods." "Where's Shanelle?" Mandi asked. "She's with my mother and Tom on a truck run," Rosie said. A tall, black Marine stepped down from the bus, eyed Rosie, and grabbed her hand. This was not her boyfriend, but she was happy to see him, though she couldn't remember his name. "Welcome home, Devil Dog," she said, and pulled him down for a kiss on the cheek. "Yeah," said the LBG. The Marine sized up the pretty young girls who were so joyful at his return. "It's good to be back," he said. "Party at the Gate," Rosie said. "And don't be late." He winked and headed for his barracks. "Hey, what's your name?" Rosie called out. Turning slightly, he said, "Val-enteen. Remember me?"

In front of the Iron Gate on Mesquite Road, several young women in hot pants and Joe Camel tank tops put the final touches on the the bar's entrance. Like every commercial establishment in town, the windows displayed gleeful homemade signs: WE LOVE OUR MARINES, THE CHAMBER OF COMMERCE SAYS SEMPER FI!, AMERICA—BACK IN THE SADDLE AGAIN. Out of the bar came Debie, wiry as always, shooting energy sparks. She wore tight pants, vest, and high heels in her favorite color, red, and her favorite fabric, leather. "Looks awesome," she said as her fellow workers tacked up red, white, and blue bunting around the door. "Okay, Jason," she called out. "We're ready." Inside the bar, Debie's thirteen-year-old son and two friends maneuvered to the entrance with a roll of thick, red carpeting and unfurled it across the sidewalk. A small fleet of Budweiser trucks approached, trailing yellow streamers, honking wildly, and stopping at the bar. "Drought's over," Debie said, and ran to greet the first driver.

The sun on this fine March day went down and the Iron Gate began to fill up the way it was supposed to before the trouble in the Gulf. As the jukebox blasted "Dirty White Boy," Marines filed in, along with the various local tribes all dressed according to affiliation, and each having to walk past Corky, who sat atop the bar at full alert in shades and a red, white, and blue doggie vest. The guest list included Crips in red bandannas; Bloods in blue; white bikers in their leathers; Samoans, who were easily identified because they looked Samoan; young girls and old women all tarted up; kids and toddlers—out here on the frontier everybody goes to a party. While the ground troops got plastered and General Norman Schwarzkopf and General Colin Powell weighed lucrative book offers, on television George Bush basked in his skyrocketing ratings: the Gulf War had been good for America, on the face of things; once again, we had control of the oil supply, and for the first time since the Korean War, it was all right, perhaps even fashionable, to be a member of the American armed services. For the military, the moment of glory was brief, and in Twentynine Palms, the Marines seized it like a beachhead.

In the back room of the packed tavern, a postwar ritual unfolded. It was familiar to locals, but this time it was more intense, carried more meaning. Underneath a sign that said, BIG BULGE CONTEST, a local band called Velvet Hammer cranked rock covers for dancers who had converged to show off their wares. About a dozen Marines of all ages and shapes, partnered with women of similar demographic and physical status, shook their hips wildly, back and forth, to the right, to the left, back and forth again, as a mob urged them on. A pair of bikini-clad Jagermeister girls circulated, pouring shooter after shooter to Marines who knelt on the floor before them, mouths wide open and bleating, waiting for their welcome-home baptism. And the drink calls echoed across the floor: "Hey, Debie, bring me a double Jack . . . Hey, Debie, I want the usual . . . Hey, Debie, Midori on the rocks . . . And give it to me nice and slow . . . Hey, Debie, I greased an I-raqi for you . . . Whachu gonna do for me?" Yes, the Big Bulge Contest was in full effect, and there was one leatherneck who danced like a sex machine, a white boy from the South, grabbing his crotch and fondling his cock Michael Jackson–style as the lead singer in Velvet Hammer belted out a liquored-up "Proud Mary." A voluptuous black girl in leather and a thicket of beaded dreadlocks jumped in front of him and mirrored his strokes, his bumps and grinds. The crowd liked this, as in "Let's get ready to rumble," and the Marine removed his sweaty T-shirt that said, GOD MADE DRUNKS SO UGLY WOMEN COULD GET LAID. He threw it to an eager teenager on the sidelines, who happened to have been wearing her own badge;

her T-shirt identified her as a member of the Lunch Box Gang, and she quickly joined in. A couple of her friends followed, dressed likewise, and then came Mandi, holding five-year-olds on each hand. "Let's show 'em our new moves," she told the kids, and now the dance floor was a mass of townies and Marines, a census report come to life only you would never find this stuff out in a census, never know how the groups were grooved together like a lock and key, all you would know is that there were a lot of kids and single mothers and Marines in Twentynine Palms. After a while, the music got slow and funky, suggesting a Gulf War victory sex show, the singer growling a mean "The House of the Rising Sun." The Marine continued stripping, removing his belt. Now another Marine cut in, pulling the dancing rasta queen his way. It was Val-en-teen.

"Turned his ass down before Saudi," said a girl who was watching as the guy showed off his big bulge, stealing the spotlight.

"What the fuck are you doing back here?" a big, burly white man called out, approaching the dance floor. He was the club bouncer, and he reminded the dancer that he was banned from the Gate. Valentine Underwood, a regular at certain bars around town, had been thrown out of the place before. The bouncer considered him a freak who was always hassling female patrons.

"It's a free country," Underwood called back.

"Tell me about it," the bouncer said. "I was in 'Nam. You didn't do jackshit in Saudi."

Underwood ignored him, continuing to grind. His dancing partner matched him, dancing more erotically, with more anger; no one was going to interrupt the moment.

"He's achin' for a breakin'," said another girl on the sidelines.

"Fucked up Saddam and thinks he's twice as bad," said another, loudly, for the benefit of the girl with dreadlocks.

"Either you go or I'm comin' in," the bouncer said. Underwood ignored him. The bouncer bulldozed through the crowd. Underwood and the girl in dreadlocks kept dancing. The white boy tried to cut back in but Underwood would not let him, so the white boy threw a punch. Underwood hit back. The bouncer—bigger than Underwood—came from behind, grabbing him by the neck. Everyone else piled on; the Big Bulge Contest erupted into war, with Mandi's two charges lost in the action. "Mom!" Mandi called as she made for the bottom of the heap, through the tangle of flying elbows and clenched fists. Debie heard her daughter above the band, the shouting, the drunks, the chorus of "oorahs" that erupted every few minutes from Marines who clanged their pitchers of beer in sloppy toasts. "Corky!" she called, and put her fingers into her mouth and whistled. Corky leaped off the counter, flying for the brawl, snapping and pawing at the bodies. "Not in my bar," Debie said as she ran with the dog into the heart of the action, diving under the mass of bodies and retrieving Mandi and the kids. In a few minutes, the party was over—the crowd had dispersed, the kids were safe, Mandi had a black eye, Debie's tip jar was overflowing with five- and ten-dollar bills, and on television, all night long, and maybe forever, there was George Bush receiving a standing ovation from Congress. As the last of the revelers headed for various desert points, the band—which included a couple of Marines who had served in World War II and Vietnam—sang the leatherneck version of "Good Night, Ladies," a parody of the original cooked up decades ago by drill sergeants at the recruiting depot in San Diego to pass on tradition to the Corpsmen of another era:

Good night, Chesty! Good night, Chesty!
Good night, Chesty—wherever you may be!

After you the Corps will roll, Corps will roll,
Corps will rooooooll,
After you the Corps will roll—on to victoreee!

Chesty was the nickname for Lewis Burwell Puller, a combat officer who was the ultimate Marine superhero, more notorious than Gimlet Eye Butler, Bigfoot Brown, or Pappy Boyington, a rough and tough bulldog who, according to legend, chased bandits in Haiti and Nicaragua, commanded the Horse Marines in Peking, battled his way from island to bloody island in the Pacific, led the landing at Inchon, and fought the most savage rearguard action in the Korean War. For a brief moment after the heady victory of the Gulf War, Marines, and through them, every citizen of America, gloried in this fine tradition, recalled with pride how the Marines had beaten the Japs at Okinawa, Tarawa, Guadalcanal, how the Marines were "the pointy tip of America's spear, out in front, kicking down the door," always the first on the battlefield, the last to leave . . . Yes, maybe duty! courage! bravery! were desirable traits after all, and so every Gulf War veteran was toasted from coast to coast, honored for his service. But several months later, after they had restored order in a region of chaos, the other war, the one at home, had resumed. This was the longest undeclared war in American history, the military war on female civilians.

A few months later, on the far side of a June evening, Valentine Underwood had an encounter with rasta queen Tammy Watson that—like so many encounters out in the Mojave—was furtive, violent, and recorded forever in a place that looked like an echo. It was dollar-drink night at Gabby's in Twentynine Palms. Tammy was living with her four-year-old son and a roommate named Rebecca in Desert Hot Springs, a nearby town known for easy availability of homemade speed—or crank as it was also called, clinics offering high colonics, and a chic spa called Two Bunch Palms which has long been a haven for depleted rich people. Tammy and Rebecca decided to drive over for the evening and take advantage of the cheap drink specials, maybe party with some Marines. There was no mistaking their high desert uniform—high heels, Joe Camel tank tops, tight jeans—for a sign that they were looking to party. Their first stop was Underwood's apartment in Two Nine. Underwood, as it turned out, was a friend of Tammy's roommate; for those who wander its paths, even the desert is a small town; eventually, one thing leads to another, and sooner or later, all travelers find themselves on the same stretch of sand. The girls parked their old red Datsun outside an apartment on Palo Verde Street. The two-bedroom pad actually belonged to a Marine named Sean who was rarely there. Other Marines spent their evenings at the pad, instead of on base, Underwood among them, calling it home, along with a small parade of Samoan locals and their girlfriends. When the girls entered, Underwood and a couple of Samoans were lying around, smoking weed. "Ladies?" Underwood offered, passing a fattie. Tammy took a couple of fast hits and recalled how much fun it had been dancing with him after Saudi. "Let's go to Gabby's," Underwood said, referring to a local bar. She passed the joint to Rebecca and left with Val in his 1984 burgundy de Ville. They got drunk fast on dollar shots of tequila, and they danced dirty to Dr. Dre and Coolio and the Sugar Hill Gang. "Hey, Val," Tammy said, "there's this guy on base, I kinda want to surprise him . . . He's in the three-eleven." "That's my unit," Underwood said. "Let's go." Outside the barracks, Tammy waited while Underwood went in on his reconnaissance mission. She lit up an extra-long Kool, took an impressive drag, and blew it out in a nice long sigh, replaying the evening. She loved to dance and tonight had been good; she hadn't been out in weeks, since dancing at the Gulf War victory party, in fact, and tonight she danced almost every dance with Val, who was really good, seemed to know all the right moves while at the same time making Tammy feel like she was the most fly chick on the floor. She hoped that the guy she liked would come down with Val; she was ready to party all night, and she looked hot. Underwood returned alone. "Hey, girl, bad news," he said, leaning into the passenger window, "your homie's in bed with some chick." Before Tammy could respond, Underwood was in the driver's seat, trashing the guy and telling Tammy not to waste her time as they headed for the base exit. "Don't worry," he said. "I'll drive you home." Tammy said that she ought to return to Gabby's to hook back up with her roommate, but Underwood again told her not to worry: "Oh, come on, Rebecca's a big girl, she can take care of herself." Tammy pondered this and, although she hadn't believed Underwood's account of what was going on in the barracks, accepted the offer of a ride because it was getting late and she needed to get back to her son. Underwood smiled and headed back into Twentynine Palms. "Desert Hot Springs is the other way," Tammy said as he made the turn. "I just wanna stop and get some liquor," Underwood said.

But he did not stop at the liquor store on Highway 62 in downtown Twentynine Palms. He kept going for about another hundred yards until they got to Utah Trail, where he turned right. Tammy was beginning to panic; this was a road that led even farther away from home, into the desert, into the official sanctuary for the Joshuas. But she did not let on that she was afraid, held on to the dear hope that he was taking a secret shortcut that somehow she had never heard of. Or maybe just his own crazy-ass way. You know how men hate it when you tell them how to get somewhere. Tammy started to fidget when he turned off the paved road onto a rutted path, heading out into the desert where in deep night, depending on the angle of the moon and the state of mind of the watcher, the Joshua trees are unwitting recipients of all manner of frightened messages, witnesses to a parade of midnight terrors, twisted wraiths frozen in time and space, keepers of all Mojave secrets. "Val," Tammy said, not sure if the words were coming out, the adrenaline racing fast now, all she could hear was her pounding heart. "I gotta get back to my son." "Don't worry," he said, "I'll give you the money for the baby-sitter." Tammy replied that she did not want his money. He stopped the car, turned off the lights, leaned over, locked the door, his eyes all crazy, eyes Tammy could peg only as windows of death. She started to scream. "Try it again and I'll kill you," he said, pinning her wrists against the car seat and prying her legs apart with his knees. Tammy stopped screaming but continued to resist: the bruise marks on her wrists would last for days.

After the rape, he climbed back into the driver's seat and zipped up his pants. Tammy was crying. "Shut up," he said. "Shut up." Tammy felt herself shrink, almost disappear, as Underwood gunned the V-8, spun the wheels hard into the Mojave, then tore across the sands and over some rocks, fleeing the scene of the crime with his sobbing victim next to him. "Shut up," he kept telling Tammy. "Shut the fuck up." Now they were back on Utah Trail, now Highway 62, turning off on Palo Verde, coming to a screeching halt on the gravel lawn of the apartment in Twentynine Palms. "Go in the house," Underwood said. Tammy was not able to move. "Do it!" The force of his voice scared her; she ran, wondering where her shoes were, somewhere during the course of the evening they had disappeared, and she fixated on her shoes, it helped her from cracking up: the gravel hurt as she ran into the party pad. She collapsed on the couch, trembling in her bare feet, trying to calm herself, waiting for a chance to run. A friend of Underwood's—Tammy noted that he was a Marine because of the regulation high-and-tight haircut—took Underwood aside in the hallway. Still flush with adrenaline, she heard their every whisper.

"What's wrong with her?" he asked.

"Her boyfriend was in bed with someone else," Underwood said.

Tammy mustered her own reply. "Your friend just raped me," she said. "How do you like that?"

"See? She's crazy," was Underwood's shrug, and Tammy bolted out the door to a phone booth around the corner at Adobe Liquors. Underwood and his friend followed. Tammy got to the phone first. She called 911. Then she called her best friend. "Please be there," Tammy said, "Oh God, Melissa, please be there." She lived nearby, on the Jehovah's Witness street, and it wouldn't take her long to come for Tammy. But Underwood and his friend pulled up in the de Ville, skidding to a halt. Tammy dropped the phone and cowered inside the booth as Underwood ran for her. "Who did you call?" he said. "Did you call the cops?" He grabbed the phone and dialed 911 himself. "Ignore the girl who just called you," he said. "It's my girlfriend and she's flipping out." He hung up and told Tammy that if she talked to the cops, he would kill her. He got back into the car and drove off. Melissa arrived in an old Accord held together with duct tape, pointing a loaded shotgun. "I'm ready," she said. "Where is he?" Long gone, Tammy explained. Melissa took her to the hospital, trying to calm her, not wanting to let Tammy know how horrified she was at the sight of bruises around her wrists. Tammy was interviewed at the hospital and then examined. The doctor concluded that her vaginal abrasions indicated either very rough sex or forcible entry. He found sperm. He tested her for gonorrhea and chlamydia. Rather than wait for the results, Tammy wanted to be treated right away. The doctor gave her Rocephin IM and a ten-day supply of tetracycline. He explained that taking the medication meant that Tammy was committing herself to an abortion, should she have gotten pregnant as a result of the incident. "Don't worry," Tammy immediately said, "I'll have the abortion." A few hours later, Valentine Underwood was questioned by local police in his barracks. "The bitch tried to pick me up at Gabby's," he said. "I didn't want to fuck that fat pig. Then she starts screamin' 'bout I raped her."

For two weeks, Tammy did not eat or sleep. She did not come out of her bedroom. Her ex-husband came and got her son. She did not say a word, just watched people coming and going from her bed through a window that looked out on the bleak and empty gravel driveway, saw concerned friends and relatives park their old Pontiacs and Chevys and get out, heard them come in the house, talk to her brother Rasta Ricky who was taking care of her, heard them say in low voices like she wasn't there: "How's Tammy? Does she need anything? She needs to see a counselor. She's flipped out. What can we do?" Nothing, she thought, not a fucking thing, and she wished they would all go away. One night, when the house was quiet and Ricky was sleeping, Tammy placed another call for help. She knew that she should have placed this call right away, but failed, didn't know how she would say the words that could end her father's career. But the desert can do that to you, in its emptiness make you connect with something other than plants that never seem to grow, make you grab on to your source and never let go, in spite of all the sunny days that can burn away all desire, even though that source was never really there when it counted. "Dad," Tammy finally said across the empty Mojave night from her rumpled bed with the tear-stained, sweaty sheets where she now passed time. "I was raped." George Watson, a sergeant major in the Marines, the highest ranking black NCO in the Corps and proud of it, was aroused from sleep in a big suburban custom-made house in North Carolina. He did not know if he had heard it right. Tammy repeated that she was raped. Then she screwed herself up again and delivered the next blow: the man who raped her was a Marine.

As Tammy would later tell friends, Sergeant Major Watson had told his daughter not to worry and asked if she had reported the incident to the police. Tammy had wept and said that she had but, she told her friends, was fearful that according to the way the Corps saw things, she had become an embarrassment, she was the kind of girl who had asked for it, she shouldn't have been hanging around the barracks, isn't that what her father had always told her? From now on she would try to be a good girl, would do as her father said, which, she inferred, was to drop the charges and let him take care of the matter. So she did, daughter of the Corps that she was, following the Marine code of "Semper Fi" as well as her duty to her father. Surely he would see to it that "things were looked into," that "something gets done," of course he would make sure that the monster was kept in his cell—he was her father. She seemed to find comfort in her memory of his words to her; she told her friends that he had mentioned something about a kind of restriction, yes, she recalled, there was to be some sort of confinement to the base until the matter was fully investigated by the Corps, and she sounded ever so slightly heartened now, the memory of her father's offering of aid tamping down the wound that would never really close now, making her feel protected and loved in a town where men were trained to respect others, to say "yes, ma'am" and "no, ma'am," a town where they learned how to give their all to strangers.

As for Sergeant Major Watson himself, he would remember the conversation with his violated daughter a bit differently. There are several kinds of restrictions, he would later tell an inquirer. If a Marine gets into trouble, he might not be permitted to leave his room. Or maybe it would be the barracks. In some cases, it would be the base. But he could not recall promising his daughter any particular kind of restriction for Valentine Underwood. Then he would sigh and say the rape of his daughter was a terrible incident, but it was a family matter, something not to be talked about with outsiders. And he would add that he was worried about what people on the base were saying about Tammy. "I heard she was hanging around the barracks," he would recall. "But this is all private, between me and my daughter."

During the years preceding the trial, Gary Bailey had warned Debie and Jessielyn in private conversations that the defense would employ a tactic that makes people hate defense lawyers. It would portray Mandi and Rosie as sluts. Although the women knew that Gary was simply trying to prepare them for what was likely to happen, neither took the news well. "Gary," Debie would say, "you're gonna have to hold me down." "Now, Debie," Gary would say, "promise me no outbursts. It won't help your case." "I know that my daughter was exploring her sexuality," Debie would say. "BUT THAT'S NOBODY'S MOTHER-FUCKING BUSINESS." Gary would agree. Jessielyn had a similar reaction, wondering why after six years of waiting for a trial, the path had led right back to where she and her sister had started—two poor country girls in the PI whom no one, not even her own government, took seriously. When John Hardy rose to make his closing argument, Debie figured the ordeal was almost over, that every abuse of her daughter had already played out. Hardy approached the jury and began. He did not know that he had already alienated many jurors by playing the race card when he suggested in opening remarks that his client had fled the crime scene after discovering the already-dead bodies because that's what black people do in America. Now he would say something that not only worsened the jury's perception of him, but narrowly missed pushing Debie into a violent act. "Why," he asked jurors, referring to his client, "would anyone have to rape these girls when the evidence has shown that they were giving it away?" As juries generally do, this one had taken great care to register no emotion during the trial. But now, a couple were clearly affected by Hardy's remark, pushed into the backs of their wooden seats by its brute force. Jessielyn rushed from the courtroom, forgetting her notepad and purse. And Debie, sitting next to Jesse Fulbright, began to hyperventilate, the heaving escalating quickly in a few seconds' time as the adrenaline rushed its course, telling her the whole thing had to be stopped now. Jesse put his big arm around her and, as he had done so often, stanched the violence of an advocate's words, soothed his charge, and prevented what surely would have been another crime from going down. Later that night, in her room at the Best Western, after having knocked back a few Jack-and-Cokes at the Jolly Roger, Debie lit three white votive candles for Mandi. A psychic had told her a few years prior to the trial that Mandi's spirit was restless and would be until the case was resolved. Knowing nothing about exactly how Mandi had been murdered, the psychic had seen the killer as a snake writhing through the lives of Mandi and her friends. A child's toy—perhaps a turtle or a little car—had lain on the floor as the snake slithered through the apartment where Mandi would die. The snake, she had said, was a myriad of forces meeting in a confluence of evil. Mandi had been trapped in its jaws. Because her death was so violent, she was having difficulty making her passage to the other side. She needed to be surrounded by white light. Debie had been lighting white candles since she had first spoken with the psychic. "I'm sorry, baby," Debie said. "I shouldn't have let that son of a bitch say those things in court today." She was not able to say the rest out loud—that she had failed Mandi again, unable this time to protect her in death. Then she paced for a while and went outside on the stairwell overlooking the freeway. It was a cold night in the desert. And windy. Cold winter wind in the Mojave doesn't feel like cold winter wind anywhere else. People say that it cuts to the bone and they are right—there is no relief and no escape; it slices like a scythe and steals your breath even as you gaze at the steam of your own weak little bellows. Chilled and tired, Debie went inside and gazed at the dwindling flames.

Defense attorney John Hardy's first witness was Ronald Bruce Gattie, a former Marine who had lived in Rosie's apartment complex at the time of the murders. On August 1, 1991, Gattie testified, he was part of an infantry battalion and was away on field duty. He returned on August 2 around 9:30 or 10:00 A.M. "My wife had left for work," he said. "I was contacted by officers later that day. I was asked if I heard anything early that morning. I said I wasn't there." But his wife was, and she had told cops that she heard a scream coming from Rosie's early that morning. "The problem with screams in the desert," Gattie said on cross-examination, "is that they could come from any number of places . . ." Hardy called Gattie's wife, Dawn Renee. She was young, slight, edgy. "I'm a little nervous," she said after she was sworn in. "Nobody's gonna bite you," Hardy said. She inhaled visibly as Hardy continued. "Where were you on August first, 1991?" he said. "I was living in Twentynine Palms," she said. "Our bedroom window was adjacent to Rosie's. We heard noise from Rosie's bedroom all the time. On August second, in the early morning, I was sleeping and I woke up to screams. The alarm was set for eight, so it was some time before that. It was a loud scream, like someone was crying. Then it stopped. I laid there for a few minutes to see if I could hear something again. I didn't. I went back to sleep. I left at nine A.M. for work." When Dawn returned from work that evening, Rosie's apartment was a crime scene. She asked her husband what had happened. She wondered if her neighbor's child Courtney had been crying that morning and went next door to inquire. Courtney's father told her that Courtney and her mother had not been home at the time of the murders. "I told a cop about the scream," she testified. "They never asked me anything else about it." Hardy concluded his examination with the observation that it was unusual for a scream to stop so abruptly, as children don't usually stop crying right away. Bailey suggested that they do if you give them a pacifier.

Inside the Joshua Tree National Park, on the way to a place called Jumbo Rocks, there was an exhibit called "The Magic Circle." The sign said:

> LOOK FOR TINY MARKS AND FOOTPRINTS IN THE SAND. COULD A POCKET MOUSE, NIGHT-SNAKE, OR COLLARED LIZARD HAVE PASSED BY? LOOK FOR HOLES—ALL SHAPES AND SIZES. PERHAPS THE DOOR TO THE SUBTERRANEAN HOME OF A NOCTURNAL WHITE-FOOTED MOUSE? THE GRAVEL-RIMMED CRATERLET OF A HARVESTER ANT? THE SHALLOW BURROW OF A BRUSH RABBIT? THE SLANTING OVAL DOOR TO THE HOME OF A SCORPION? LOOK FOR COLOR. THAT GRAYISH SHRUB IS ACTUALLY A SHADE OF GREEN. THAT ANIMATED BIT OF SCARLET PLUSH IS A LITTLE HARVEST MITE. YOU'LL HAVE TO GET DOWN ON YOUR STOMACH TO ENJOY THE TINY PINKISH FLOWERS OF THE DESERT CALICO AND THE CUPPED BLOSSOMS OF THE PURPLE MAT. LOOK FOR MOVEMENT. GROUND BEES MOVE IN AND OUT OF THEIR TINY BURROWS. WRIGGLING ITS FLATTENED BODY FROM SIDE TO SIDE A HORNED LIZARD DISAPPEARS HEAD-FIRST INTO A BLANKET OF SAND. A JACKRABBIT BOLTS FROM BENEATH A GRAY-GREEN SHRUB. LISTEN. A FLY DRONES. A SMALL GREEN CREOSOTE GRASSHOPPER CHIRPS. A THRASHER BURSTS INTO SONG . . . THE MAGIC CIRCLE. WALK OUT INTO THE DESERT A LITTLE WAY. IN YOUR MIND'S EYE DRAW A CIRCLE AROUND YOURSELF, OUT ABOUT A HUNDRED FEET. NOW, LOOK FOR LIFE IN YOUR MAGIC

CIRCLE. WHOEVER SAID THAT DESERTS ARE PLACES OF DESOLATION? LOOK AT ALL THE LIFE IN YOUR MAGIC CIRCLE!

Mandi Scott was not big on hiking, but she had her own magic circle in the desert. Her great curiosity about people and passion for life had taken her on a journey away from her world, into subterranean homes and burrows and craterlets where she was drawn by the sound and the color, but blinded and deafened to the danger. In fact, during Debie's pregnancy with her second child, doctors predicted that Mandi was a boy because the heartbeat was so fast. It has been said that the act is the decision, rather than the act of the decision-making itself, and from an early age, Mandi was always acting, always going, always doing. There was the time that she tried chewing tobacco at five and got sick. There was the time that she was playing with Jason and Krisinda and she ran out the front door and was hit by a car. Debie heard the awful sounds of the screeching brakes and rushed from the kitchen to the street. The car was dented, and later, a doctor told Debie not to worry: "She's got a sore arm and a little road burn, that's all." There was the time that the three kids were watching Debie ride a bucking bronc in a rodeo. When Debie got bucked off, eight-year-old Mandi ran to her mother with a Bud and said, "Here, Mom, have a beer." That was the first of many times that Mandi offered her mother such assistance, and over the years, the moment assumed official status as a funny family story. In fact, Debie had told the story to many of her customers at work, and even those who had never met Mandi knew that she was "good people."

Although in the course of ten years, Debie would switch jobs five times—tending bar at the Staff Club on the Marine base, Pepper's, the Catch 29, at the Best Western Motel, the Iron Gate, and the Oasis—the transient family had finally stopped moving, sinking roots in this parched and sandy region as easily and quickly as jumping cholla carried on the wind. At Twentynine Palms Elementary School, a group of girls immediately came together around outgoing Mandi. Although of different races and ethnic backgrounds, the girls had one thing in common: they were offspring of the working poor and generally lived with their mothers or other female relatives.

During their elementary-school years, the girls were inseparable; they had birthday parties at Chuck E. Cheese, occasionally went with an older brother who had a driver's license to the beach in San Diego, baked cupcakes and decorated them with M&M's, got kicked out of class for being rowdy. All average students with gifts that were largely unrecognized outside their circle, members of the group formed an alliance that was fierce and intense: they would engage in physical fights to defend each other; if one had an enemy, so did they all. By the time they got to junior high, Lydia, Tina, Tam-me, and Janita were having lunch every day at Mandi's, which was across the street from school (on Wildcat Drive, a reference to the school teams, and also, to the delight of Mandi and Debie, their own zodiac sign). In a town rife with gangs, Debie nicknamed the girls "The Lunch Box Gang." Soon they had T-shirts printed up, and the LBG, at least in terms of presence, had joined the ranks of the Watergate Crips (a branch of the black L.A. gang), the Park Village Crips (a part of the Samoan gang from Compton), and the local Bloods (also from L.A.).

By the time Mandi was fourteen, she had developed a tight connection with both sets of local Crips. The black Watergate Crips were kids, mostly boys, some of whom had been shipped to Twentynine Palms by the Los Angeles juvenile court system as the first part of a choice—the desert or jail. One of the first to arrive was Kevin James. It was 1989—the height of the crack epidemic in Southern California—and he was fifteen. His mother, just twice his age, was a crack addict who lived in nearby San Bernardino. Kevin had relocated to a low-income housing section of Twentynine Palms with his grandfather, younger brother, and various cousins. The plan was to stay clean and straight. The tall and skinny Kevin—or "Bone"—had already been in and out of juvenile detention for possession of pot and driving without a license.

On August 4, 1990, Mandi turned fifteen. Debie made Mandi her favorite dinner—spaghetti and meatballs, followed by a refrigerator cake, topped with fifteen candles. Mandi tried to blow them out and got all but one. Kristy and Jason clapped and demanded a speech, but Mandi burst into tears and ran to her room. For the rest of the night, she stayed there, speaking to no one unless the phone rang. "Is it for me?" she would say. Debie would call out no. "Last call for cake," Debie said as she plucked the candles and put the plate back into the refrigerator. Still, Mandi did not answer. The next day was in every way a repeat of August 4, except it wasn't Mandi's birthday: the phone rang, Mandi inquired, and then her flashing eyes lost a little bit of their famous green luster when she learned the call was not for her. On August 7, Debie made the call that Mandi had hoped would come to her. "Hello, Max?" Debie said, and explained how Mandi had waited all day on her birthday for her father to call. Max told Debie he had forgotten, and asked if he could speak to Mandi. "Dad," Mandi said, her voice quavering, "what happened?" Max apologized, and said that on the big day, he was at Six Flags Over Texas with Kathy and the kids—"you know how much I love that roller coaster"—and had lost all track of time. Mandi told her father that she hoped he had fun, hung up the phone, and then disappeared into herself, reappearing over the following months when she was needed, which was often, fighting everyone's fights and nurturing all the children of children, but with a flame that only her mother knew was diminished, as if someone had lowered the wick in a lantern.

A few weeks later, Mandi started tenth grade. Her favorite classes were choir, gym, and English. She was a natural at the first two, and struggled impressively at the third, taking to the study of American poets such as Emily Dickinson and Edgar Lee Masters. In particular, she liked a poem by Masters that took place at the Malamute Café. One day she boasted to her friends: "I know what a malamute is, do you?" Then she explained to them the difference between a malamute and a pit bull, in great detail. "Pit bulls have twelve or thirteen per litter, malamutes only have nine or ten." "Maybe you should be a veterinarian when we get out of here," Tina said. "I don't know," Mandi replied. "Do you have to go to school every day?" Although she had her favorite classes, it just was not in Mandi's nature to actually get to school every day. "Please excuse Mandi's absence," she would write in Debie's handwriting, "as she has the flu." One day, the attendance counselor called Mandi to his office. There were five or six students waiting to see him, all in the fashion of the day—baggy pants and tank tops, partly laced Nikes. "Hello, young lady," a tall, good-looking black kid said to Mandi. "Have a seat." "That's okay," Mandi said, but he was already up and out. "Thanks," she said, and took his place. "What'd they call you in for?" he said. "Cutting class," Mandi answered. "What about you?" "Wearing gang attire," he said. The kids started laughing—gang attire was basically the school uniform, and they were all wearing it. The counselor peeked out of his office. "Miss Scott, would you come in?" "Does Miss Scott have a first name?" the black kid said. "Mandi," she said as she got up. "I'm Kevin," he said, extending his hand. "Kevin James. They call me Bone." "Miss Scott," the counselor said, "this is not the time for meeting and greeting." "Thanks for the seat," Mandi said, and then added flirtatiously, "Bone." "Catch you later . . . Mandi," he said as the counselor guided Mandi inside and shut the door, and then to the other kids: "I will, too."

Like many girls on the edge of nowhere, Mandi was starting early. Before Kevin, there had been other boyfriends, but now it was true love. Every day after school, Mandi went to Kevin's house, and she was there on Saturdays and Sundays, walking the three miles from home and getting a ride in the evening from Kevin or one of his friends to her baby-sitting jobs around town. Mandi liked it at Kevin's. His cousins and their girlfriends were always there, welcoming her into the tribe ("Yo, Miss Scott, how ye be?"; "Hey, girlfriend, lookin' good!"), and Marines often dropped by, buying bags of pot or vials of cocaine from Kevin and his fellow Crips, bringing the latest CDs from the exchange on base. The Marines were always so polite, Mandi thought, she liked it when they said "yes, ma'am" and "no, ma'am," felt like she was important in a town that was pretty crude, even if that was how Marines were trained to talk. And she liked that they asked her out and that Kevin was there to say that she was taken. "She's all mine," he said when he heard a boot ask how old she was. "Right?" Mandi nodded and kissed him. "Right," she said. "Very right." And that's how their daily make-out sessions started, triggered by Kevin's jealousy and possessiveness and the urge to make a display, and fueled by Mandi's need to be protected and desired. "You are so beautiful, baby," Kevin told her as he took her to the couch. "How would you like to have a zebra baby someday?" "My mom would kill me," Mandi always said. "I can't." "Don't you wanna have a little Mandi?" Kevin asked. "Raise it up with me? We could go to L.A., stay with some of my homies. Then we could have a whole big zebra family. You be the mama zebra, I be papa." Mandi thought about this image and started to smile, but caught herself. "Then my mom would really freak," she said. "You leave your mom to me," Kevin said. "Ain't no lady can resist the charms of Bone. Am I right?" He held her tight for a long kiss. "Baby, you got some nice curves," he said. "I don't want no one else touching them, you hear?" Mandi promised Bone that she would never be with anyone else, and he said that he would never leave her. As the pair exchanged vows of fidelity, the young Marines did a couple of hits of rock and then shacked up with girls at the house or, if none were left, went elsewhere, looking for love at other desert dens. And Tone Loc rapped on, the voice of sex itself, urging the planet to do the wild thing.

In April of 1991, Kevin broke up with Mandi and took up with Janita, a black member of the Lunch Box Gang. Soon he was sent back to jail on a DUI. In her secret journal, Mandi wrote a poem called "The Mystical Man":

There is a mystical man in my thoughts.
And in my dreams he is always there for me when I scream.
Please mystical man tell me why you're here.
Have you come to dry my tears?
Have you come to save me once again?
Or are you trying to save me from him?

She sent letters to Kevin. "Dear Bone," she wrote. "I don't want to live anymore. If I can't have you, there's nothing for me. Every day I think about cutting my wrists." He got word to her through a cousin that he would see her when he got out. "Tell Mandi not to go crazy on me," Kevin said. "Tell her to hang in. Be strong." But every day at school, Mandi watched Janita's stomach grow. She was carrying Bone's baby, and when he got out, Janita said, they were going to get married.

One winter day, deep in the heart of sixth grade, Mandi and her close friend Tina were walking home from school, across the sandy expanse between Twentynine Palms Elementary and Baseline Drive. "I think I have to go to Arizona to live with my father," Tina said. "Why?" Mandi asked. "I don't know. My mom wants me to." "I can't live here if you're gone," Mandi said. "Guess what this says?" she continued, making a fist and holding it up so Tina could read the letters she had written on each knuckle. "I—P—M—A—T . . . ipmat—what's that?" Tina said. "I don't know," Mandi said. "I thought it would help me remember the phases of cell division during the test but it didn't." "I skipped that question," Tina said. The girls usually walked past Baseline and then south toward home, but now they continued on a westward route, stopping at the church on the Kentucky Fried Chicken road and peering through its dusty windows, then continuing on, lost in the sad news of Tina's imminent departure, west toward the no-man's-land that lurked just outside of town, where the only hint of civilization for the foot traveler was the occasional jetliner that streaked the twilight for Los Angeles, or the carcass of an abandoned vehicle, like the old county bus that Mandi spotted way out on the horizon. "Let's go live there," Tina said. Mandi said that they didn't have underwear or sweaters, but Tina thought they would be okay—"I learned how to make a fire in Girl Scouts."

But as twilight faded to night and the coyotes struck up their chorus, the girls lost sight of the phantom bus. A moon sliver provided a hint of light which illumined the hulk every now and then, but an endless parade of passing clouds rendered this guide useless. Mandi announced that they were lost after she had failed to sight the bus for several hours. "Are you cold?" she asked Tina. "Here." She slipped her arm out of a sleeve of her big Wildcats sweatshirt and beckoned Tina to wriggle under the empty half. The two girls lay down, head to head under the improvised blanket, in the accommodating, endless arms of the Mojave. They watched comets and their body heat was trapped and protected them against the gathering cold, and they talked about moving to Santa Cruz after high school. Tina's grandmother lived there, right near the ocean, and they longed for water and waves and for the chance to swim with dolphins. "I read about this girl who kissed a dolphin," Tina said. "No way," Mandi said, and then asked Tina if she could visit when she moved to Arizona. For a moment Tina didn't respond, then looked away and told her best friend that she didn't think people from Twentynine Palms were supposed to visit her there.

Mandi thought about this for a while, wondered how come she didn't know anybody whose parents were married, how come all her friends had fathers who were somewhere else, now feeling heavy and tired under the chilly night sky with a particular memory of something terrible that had happened when she was six years old. It was time to tell Tina. Mandi began the story about the Christmas when her father told her that she would never see her mother again. She and Jason and Kristy were visiting Max in Midland. On Christmas morning, as they unwrapped presents, Jason said that he missed his mother and wondered when they would be going home. "Your mother is dead," Max said. "You're not going anywhere." Jason and Mandi started crying and Kristy tried to calm them, guessing that Max was just making up a story. But Max insisted that Debie had been killed in a car accident. Several days after Christmas, the cops barged into the house and took the three children to a motel near the airport, where Debie was waiting in a room. When Jason saw Debie, he screamed; he thought his mother was a ghost. Mandi held him until the shock had passed. That night, fearing that Max would try to reclaim the kids, Debie barricaded the door with furniture. They finally fell asleep on the king-size bed. Early the next morning, the cops escorted Debie and her kids to a plane and the family returned to California. "My dad was arrested for not sending us home when he was supposed to," Mandi said. "But he got out of jail in three days."

From a distance came the sweep of big headlights. "Guess it's time to go home," Tina said. "I hope you don't have to move," Mandi said. "Me, too," Tina told her, "but if I do we'll always be best friends." Caught in the glare, the girls looked like conjoined twins, both with long hair—two heads atop one body—did this freak live out here?—but Debie knew it was Mandi and Tina, stranded. She ran to the girls with Jason and several cops, and told them that she thought they were dead and lying in a ditch somewhere. A cop said that they gave everyone a big scare, warning them who knows what kind of crazy people live out here, they might have gotten cut up and put in an ice chest. "Lots of weird things happen in the desert," he told them.

PART THREE

Lambs to the Slaughter

When an evil spirit comes out of a man, it goes through arid places seeking rest and does not find it.

Matthew 12:43

Take a look at me
Tell me do you like what you see
Do you think you can
Do you think you can do me . . .
Girl let your hair down
Take off your clothes
And leave on your shoes
Would you mind if I look at you for a moment
Before I make sweet love
Do me baby
I like it in the morning time
Do me baby
Sometimes I love it in the evening.

Bell Biv Devoe, "Do Me"

Two weeks after the murder at the Texaco, Rosie could still see the victim's blood on the asphalt. Maybe it was a good sign, she told herself. Maybe this was the murder her mother had foreseen, which meant that she was off the hook. Still, the sight of the blood every day was causing nightmares; she hoped she did not have to continue working at the mini-mart for much longer, hoped that the relationship that she had with a Marine in the 3/11, Lance Corporal Timothy Carmichael, would get back on track. They had met after the Gulf War and had fallen in love. But Rosie couldn't stand being alone, especially after the murder. One night, when Tim was out on the rifle range, Rosie took up with another Marine. Tim broke up with Rosie when he found out about the affair. Out of loyalty to a friend and fellow boot, Rosie's new boyfriend dumped her.

On the morning of July Fourth, Rosie took Shanelle to the base to hear the Marine marching band ring in the holiday. The ceremony on Condor Field was elaborate and evocative:

Marines from various battalions wore uniforms from all periods of the Corps' history, from the days of the colonial wars through the world wars, Vietnam, and current conflicts. When the band struck up "The Marine Hymn," a few of the older veterans in the crowd sang the rousing call to arms about the halls of Montezuma, the shores of Tripoli. Rosie and Shanelle clapped along to the beat. A Filipina sitting in the bleachers with her young Marine husband turned to Rosie and asked if she ever got homesick. Rosie said that she missed her grandparents, but not the PI. "Hey," Rosie added, responding to the woman's longing for familiar terrain, "I'm having a party, why don't you come over?"

In the early afternoon, back at her apartment, Rosie prepared for the day's guests, removing trays of her homemade lumpia and dishes filled with rice and cinnamon milk from the refrigerator and placing them on her dining-room table. Mandi and Lydia came over to help her get ready for the big bash. Lydia brought tamales that her mother had made. Mandi brought a case of Dr Pepper and several jumbo bags of M&M's that she had purchased with her baby-sitting money. M&M's were her trademark snack, ever since her father forgot her fifteenth birthday. From that moment on, she told everyone her last name was McMaster, and began signing schoolroom notes and love letters with the initials M.M. Out here, in the land of disconnect and the quick study, brand-name affiliations confer and convey an immediate identity; it wasn't long before everyone in Mandi's magic circle knew that her candy of preference was a badge. Rosie handed Mandi three red, white, and blue china bowls she bought for the party at the Pic 'N Save in Yucca Valley. "Here," she said. "The best for the best." Mandi filled the bowls with the M&M's and arranged them around the apartment.

Guests began to arrive in the late afternoon, as the festivities on base were winding down. Among the first was Jason, who followed Mandi everywhere. Mandi liked to have Jason by her side, and this urge had become more keen ever since Max Scott had tried to keep them. Tonight was one of those nights when she needed Jason more than ever. She was still pining for Kevin James, although strangers would never have guessed: the best dancer in the house, Mandi was busting some moves that even her dark-skinned homies had to admire. Rosie was a close second, rocking up a storm with her daughter and anyone else who wanted to jump in. By sundown, the tiny apartment was packed, all tribes represented, and then some—word had gotten out as far as Amboy, a pit stop eighty miles to the east on Highway 62. As firecrackers exploded all over the desert, the country, American military bases all over the world, boys and girls started to pair up. But there were not enough girls to go around, and not every girl wanted to pair up. This was not good news for a couple of Marines from the 3/11, among them Valentine Underwood. As their advances became more aggressive, a couple of local boys took offense on behalf of the girls. Things were said, punches were exchanged, knives were flashed, the cops were called—it was the 215th birthday of America and the people who fight the country's wars spilled a few more drops of blood in the defense of the pursuit of happiness.

As the military saying goes, Marines come into town to stir up civilian unrest. As townies say, the cops don't do shit because they don't want a hassle with the Corps. And men are idiots; that's what girls everywhere say, and that's what Rosie, Mandi, Lydia, and a few other members of the LBG said as they cleaned up after the party. Later, the gang sat down to play some blackjack. Mandi hoped that Kevin would come back to her, in time for her sixteenth birthday next month. "Do you think he remembers?" she asked everyone, no one in particular. Her friends assured her that he did, hoping that saying it would make it so, but knowing that with Janita now pregnant, a far more likely scenario would be for Kevin to get himself busted again. Rosie hoped that she and Tim would make up soon. She shouldn't have slept with his homey. "Maybe he doesn't believe me that it was no biggy," Rosie said. Her friends assured her that in time, he would forgive her. It takes men a long time to get over these things, longer than women, they told her. The girls all spent the night at Rosie's, and early the next morning, they headed home, all except Mandi, who planned to stay for the day to take care of Shanelle while Rosie went to work. It was a great party, they said. Great food! You did good! Everyone loved it! And they did, this evening that as far as the Mojave was concerned was just one more desperate interaction played out on its flanks, one more grab for connection, another sorry attempt to sink some roots before it reclaimed the whole thing forever, anyway. Just another goddamn party where some asshole calls the black-and-whites. Like that's gonna change anything.

Rosie asked her daughter, "Who loves you the most?" Shanelle pointed to her mother and said, "You do." "That's right," Rosie replied, "I'm the one," and then she picked up Shanelle for a big hug. "See ya later!" she said, heading up Palo Verde to the Texaco. "See ya later!" Shanelle said. "Don't forget to call me," Rosie said, gesturing as if she were on the phone. "I won't," Shanelle said. This was a little game they played; Rosie had given her daughter a toy phone to make special calls. "Bye-bye . . ." Rosie faded out, disappearing in the blinding white light of the desert morning. "Bye," called Mandi and Shanelle. July 5, same old, Rosie thought as she rounded the corner on Highway 62. "Heigh-ho, heigh-ho," she started singing, "it's off to work I go . . . I'm so bent, need the rent, heigh-ho . . ." It was a local bar joke, and every laborer in town was singing it.

That afternoon, a Crip stopped at the Texaco. "Hey, Fish," he said, rolling down the smoked-out window on his regulation Mercury Monte Carlo. Rosie took it as a term of endearment.

"I'm running on E and I ain't got no money." Rosie peeked out of the glass booth. "That's okay, homes," she called, making sure to be heard above an ambulance that was screaming toward another DUI wreck at a nearby intersection. "Pay me when you have it." She turned on the switch for pump number three and winked at the Crip.

Rosie's boyfriend Tim discovered the bodies eight hours after the murders, in the early evening. He and a Marine buddy named William Short had pulled up in front of Rosie's apartment. He was happy to be off the rifle range and looking forward to getting back together with Rosie. He wanted to tell her how much he loved her, and that he was no longer upset about her affair. He left the car and entered the unlocked apartment with the anticipation of putting things right: on the rifle range, he had decided that he wanted to get married. There was static coming from the radio and he thought that was strange. "Rosie," he called, "baby, you here?" Just more static, and the low drone of the swamp cooler. The living room was empty, so he headed toward the bedroom. He saw Mandi lying on the floor, quite still. She's naked, he thought, and approached cautiously to see why she was not wearing any clothes. He saw holes in her chest and could not understand what they meant. I should ask her why she's lying there with no clothes on, he thought. Maybe she needs a blanket. But his eyes drifted and he saw Rosie. She looks just like Mandi, Tim thought. She was naked and on her back, her body splayed across the entrance to the bathroom, with her head outside the bathroom on the shag carpet and her feet on the tile inside, next to the toilet. He noticed that Rosie, too, had holes in her chest. The same kind. I should ask the girls what happened, he thought. But he didn't. He knew he had come upon an ambush of some kind, and he turned and left, motioning for William to come into the apartment, unable to say anything other than, "Hey." William followed Tim, stepping over Rosie's body, into the bedroom, looked at Mandi, the knife, the blood, then at Tim. "Don't touch nothing," he said. The two walked quickly out the door. They had seen dead bodies in Saudi but nothing like this. They went down the street to the Samoans' apartment and asked to use the phone. It had been disconnected because someone had forgotten to pay the bill. They proceeded to another friend's and dialed 911. "You weren't afraid to call the police, were you?" Bailey asked Carmichael. "No, sir," said the dark-skinned black man.

A few days after Carmichael's testimony, Jessielyn celebrated her thirtieth birthday during happy hour in the Jolly Roger at the Best Western. She ordered a purple passion and one or two sips later was in tears. What upset her wasn't so much turning thirty as outliving her younger sister. "Rosalie would have wanted me to have a good time," she would say as she tried to stop herself from crying. But then the tears would flow again. "You can't let Underwood get to you like this," Debie would say, as she often did. "You are never going to get through this trial and it's only day three." Jessielyn asked Debie if she could borrow some strength.

July 28 was Debie McMaster's birthday, her thirty-eighth. Never one to miss a chance to party, she invited a few friends to the Iron Gate. Mandi came, bringing the LBG, and gave her mother a stuffed tiger. But the chatter was more about another birthday, Mandi's, which was coming up on August 4. It was her sixteenth, and Debie was planning a big surprise. Since the boys had come back from the Gulf, she had been able to stockpile enough money in tips to make the final payment on a 1980 Camaro, sky blue, which she was planning to present to her daughter on the big morning. A powerful car for powerless people, the Camaro was often seen cruising the streets of Twentynine Palms, desert towns across the land, and in fact any town where people were mostly on the skids; possibly it was the vehicle that spawned the Mojave mantra, "Eat my dust." Debie had been hinting about the big surprise for months, not just to Mandi but to all of the family's friends. She was hoping that the car would be more than just a rite of passage for her teenage girl, that it would be Mandi's ticket to ride—out of this place, out of this life, away from gangbangers and the military and all varieties of trapped lowlife to a faraway land that her family's legacy could not penetrate. But something was wrong; it seemed that the only person in town not looking forward to her birthday was Mandi.

Mandi was still depressed over her breakup with Bone. And there was the likelihood that her father would again forget to call, coming up with another lame excuse. Mandi dreaded August 4. The day after Debie's birthday, Mandi began checking in, one by one, with her buddies in the LBG. She told them she did not want to have a party. The girls were very disappointed, most of all Tina, who never did move to Arizona and, true to the town's code, was never asked to explain why. She and Mandi shared birthdays; Tina's sixteenth was two days after Mandi's, and the girls had been planning to make it a double celebration from the moment a few months earlier when Tina had learned she would not be moving. As her birthday approached, Mandi began spending more time away from home, away from the telephone that did not have Bone on the other end, hanging around with some girls that others in the LBG did not like because their preferred male companionship consisted of Marines, who, although fun and more regularly flush with cash, were generally regarded as more trouble than the local gangbangers. Among those with whom she maintained contact was Rosalie Ortega.

During the last few days of July, Mandi was a frequent visitor at Rosie's, stopping by to check on Shanelle, baby-sitting one night when Rosie worked the night shift. Like Mandi, Rosie was feeling blue. She and love-of-her-life Tim were in touch again, and there was talk of getting back together, but so far the contact had been primarily by phone. During the last week of July, just when the moment seemed right for a face-to-face reunion, the Corps scheduled Tim for time on the rifle range. Rosie felt anxious, had the feeling that things were not meant to be.

On July 31, Juanita and Tom Brown picked up Shanelle for a truck run to Las Vegas. They would be back in five days. Rosie handed Shanelle her little suitcase and her toy phone and asked her who loves her the most. "You do," said Shanelle, kissing her mother on the cheek. "You call me from the road," Rosie said. Juanita nodded that they would. With a final hug, Shanelle skipped away as she always did when embarking on these trips, happily kicking up stones in the driveway, calling good-byes to her mom, and then waving to Aunt Mandi, joining her grandparents as they climbed into the Peterbilt and headed away from Twentynine Palms.

On March 6, 1998, Department D, the Victorville courtroom of presiding Judge Rufus T. Yent, was filled with all of the surviving members of Mandi Scott and Rosalie Ortega's families. Although a verdict had been handed down on December 19, 1997, sentencing was delayed until the following spring. Once again, Debie McMaster had flown back to the desert from Chicago, along with her longtime companion, Mike Ramirez, who had left Twentynine Palms years ago and had wanted never to return to the Mojave. Jason had returned with them, and they were met by Krisinda, now working as a veterinary technician and living with a Marine outside San Diego. Rosie's mother Juanita

Brown had come to court for the first time, driving in from Las Vegas with ex-Marine Tom Brown and Shanelle, now a nine-year-old schoolgirl learning to dance "The Nutcracker" and play the violin. Jessielyn had driven up from Camp Pendleton, expressing her husband's regrets that he could not join the family; he was on duty. Also on hand was the ever-beleaguered Jesse Fulbright, shuttling between cases in Barstow and Joshua Tree, and Mandi's friend Lydia Flores, who had taken a day off from her job as a clerk at a cable television station to drive in from nearby Apple Valley. No one from the defendant's family was present. But there were nine cops, in addition to the three court bailiffs; evidently there had been concern that Debie might lose her temper and take matters into her own hands Sonora-style. There were also various members of the district attorney's staff, relieved that the case was finally wrapping up and moving out of Victorville, vanishing into the desert which had called it to life.

The defendant, in his orange high-security jumpsuit and ankle chains, was escorted into the empty jury box, where he faced the gallery. In a few moments, he would be en route to a new home, Tehachapi State Prison, on the 58 south of Bakersfield, and then several months later, Corcoran State Prison, on State Highway 43, midway between Bakersfield and Fresno, a few hours from Sonora, where for the rest of his life his neighbors would be Sirhan Sirhan and Charles Manson. It was all too fitting, too perfectly grisly, that Underwood would live out his days in the western lowlands outside the Sierras, a bizarre landscape marker of what happened to the women in Debie's family whenever they came down from the mountains to find a better life. Judge Yent asked the victims if they had anything they would like to say to the man who had been convicted of murdering daughter, sister, friend, loved one.

Shanelle rose and made her way through her row of spectators, squeezing past Jesse Fulbright on the aisle, and walking with the poise of a student of ballet—or a child who has grown up too quickly—to the podium. She had typed her statement on her computer, and from the confidence with which she read aloud, she had been waiting and preparing to do this for a long time. The convicted killer seemed to conjure a spirit of boredom as the little girl told him of how life had been since the day she lost her mother.

> *I remember me driving with my mama and papa in my papa's truck, on our way to Mommy's house. On our way there my mama taught me a trick, look there is something on your shirt, then you would look and I would say "got ya." I asked my mama if I can show that trick to my mommy when we got home and she said of course. When we got there we saw people crowded around my mom's house. I said to my mama that I think something could be wrong with my mommy. My mama told me to shut my mouth. My mama started to talk to someone and then she started to go insane. I kept on asking my mama, "Where is my mommy, Mama, where is my mommy?" but she would not stop yelling and kicking. I had no idea what was going on because no one told me until five to ten minutes had passed. I did not do anything. I did not know what to do, was I just suppose [sic] to cry right there? I wanted to cry, but I just could not cry.*
>
> *After a while later I started to cry everytime [sic] I thought about her and whenever someone talks about death. I get scared everynight [sic] when I am about to go to sleep because I will never know what is going to happen. Someone could just brake [sic] in and kill me or hurt me, I am scared to watch any scary movies that have anything to do with stabbing. I never thought that any one of my friends or family members would*

> *die so soon, but I guess I was wrong. At least let her die for a reason because I don't see why people would do such a horroble [sic] thing to an innocent person. He did it for no reason. He is just so stupid. Why would he want to ruin someone elses [sic] life.*
>
> *I don't know that much about my mommy because I have only known her for 5 years. Sometimes I would have to go [to] my mama's house because my mommy had to go to work. I wish I could spend more time with her. Every time I win an award or get strait [sic] A's, I really wish that my mom was there to see me. I know she is there in spirit, but that is just not good enough.*

On her last night, Rosie asked Mandi if she would stay with her until Tim got back from the range: "Homes, I can't be alone." Mandi agreed; keeping Rosie company, adding a warm body to the empty apartment, might take the sting out of her approaching milestone, make her feel like she counted. Being a friend was what Mandi did best, and for the first time in weeks, she felt a little bit lighter.

It wouldn't be the first time Mandi looked after people who were older than she, especially those whose families were fractured. When a married lance corporal she had met at the Oasis was away in Okinawa, she spent hours with his toddler. His wife was having affairs with other Marines, and Mandi was concerned that the child wasn't getting enough attention. She couldn't watch and pretend that nothing was going on when she knew that here was yet one more family shattering before her eyes. "I hate to tell on people but I think there's something you should know," she finally wrote to her friend in Japan. "Your wife is sleeping around and it is with people in the Corps." The lance corporal boarded the next States-bound plane as soon as he received Mandi's letter. But it was too late to save his marriage, and his career. Busted for adultery and thrown out of the Corps as a result of a rule which proscribed extramarital affairs for Marines and their spouses, he joined a strange and haunted legion of desert wraiths; they come and go at twilight and all through the night, shuffling in solitaire across the empty streets from bar to bar until closing time, some carrying pool cues in special cases, others with nothing to grab on to, gorging on drink specials and wondering what it was that they did wrong, how it was that the laws of gravity did not keep them in another place, exactly what the fuck happened to cause this weird and unmarked beaching of certain people in Twentynine Palms?

In the heat of a desert summer, everything is still, trying to conserve water. Everything is waiting, as everything always is in the desert. But in the summer it's different. The wait actually has a goal: it's the passage of time. As the sun reaches and begins to descend from its peak, every granule of sand through the hourglass means the temperature is that much cooler. At night, things get crazy—with air circulation and a darkening sky, the furnace door swings open and everyone and everything makes for the exit in search of something: snakes uncoil from their creosote nests and slide out to hunt; kit foxes emerge from their burrows; plants with spikes and ridges release heat and expand ever so slightly; Marines change clothes and flee the barracks; kids page other kids, trying to locate the next party.

Mandi and her friend Beth spent the night of July 31 at Rosie's. Also camped at the Ortega fire for the evening were a couple of other LBG members and a few Marines. The next morning, Rosie went to work, and Mandi accompanied Beth back to the base where she lived with her mother and stepfather. Although civilian girls were not permitted to hang around the barracks, this rule was often overlooked, and the two friends spent the day in and around the 3/11, flirting with Marines who were off duty, wandering the hallways, commenting on the music blasting from certain rooms. In the late afternoon, Beth spotted an MP she knew. "Hey, Bubba, wanna party?" she said, knowing he couldn't, not yet anyway, and he told her, no, ma'am, and kept walking. Beth suggested to Mandi that they party with him when his shift ended, trying to convince Mandi to wait by pointing out that the MP had a black roommate who had some really good pot. Mandi begged off, explaining that she was not in the mood, but Beth persisted. "Look," Mandi said. "I'm just getting tired of this whole thing." "Great," Beth replied. "Do what you want. You'll never get over Bone that way." "Fuck you," Mandi said, and walked off.

At a pay phone near the commissary, she called Lydia and said she was on her way to Rosie's. "Wanna hang out?" Lydia replied that she couldn't; she and a family friend had been "messing around" and she broke her arm. "I just got back from the hospital," she said. Mandi took this in stride, another Twentynine Palms mishap that probably had another story. Mandi told Lydia that she hoped she was okay and that she would see her the next day at the Silver Quarter, a video arcade that the LBG favored. As she walked toward the gate, she thought about how Lydia sounded. She knew the part about Lydia breaking her arm was true, but she wasn't sure about the rest. Did somebody purposely hurt her? She seemed so tough, but you could never tell in this town. Once, when Lydia came to school with a black eye, Mandi wanted to ask her what really happened, but didn't, mad at herself for becoming just another townie following the Mojave code of mind your own business, a way of life coined long ago for the benefit of all who walked its sands. Was it like this everywhere? Mandi wondered as a caravan of tanks commandeered by teenage boys passed her, on their way to or coming back from war games in the desert. Would she ever find out? Probably not, she thought as a couple of young boots in a Ford pickup saw her and slowed down, called "Hey, Mama," and then, ignored, called again: "You're kidding, right?" Mandi let it slide and the Marines reluctantly moved on. This was her life, her crazy life, and she could handle it.

Just outside the gate she ran into a friend who worked on the base and was heading back to town after her double shift at the PX. "Sure, I'll drive you," the woman replied, happy to encounter a member of her extended family of the working poor; such tasks as offering rides kept the group together and permitted grown-ups to keep an eye on kids without violating all manner of unspoken understandings. "Jeez," the woman said, puffing on a Marlboro as Mandi got in, "this place is driving me nuts." "Tell me about it," Mandi said as the two friends, young and middle-aged, shared a laugh. The older woman's laugh was laced with a lifetime of tar and nicotine. At the end of the laugh, she coughed up some phlegm, swallowed it, and sucked on the coffin nail again. "Wish I could get off this stuff," she said, "but nothing works. Keeps me sane, I guess." As they approached the intersection of Adobe Road and Highway 62, Mandi fished a cigarette out of a pack on the floor, lit up, and took a drag like a pro. "Easy, girl," said the friend. "You got plenty of time." "I know, I know," Mandi said, then, at Palo Verde, told her friend to make a left turn: "I'm going to Rosie's." The friend remarked that she liked Rosie very much. "Whenever I gas up at the Texaco," she said, "she always tells me I look good."

It was about five or six in the evening, very bright, the white sands refracting cleanly off the high-angled summer sun. Mandi walked up the driveway to Rosie's. A desert blow was coming in; the town was breathing again, emerging from the day's stillness. Rosie heard the crunch of Mandi's high-tops on the gravel, a comforting sound—company coming—and opened the squeaky screen door. As Mandi entered Rosie's, she turned and waved to her friend, who was waiting in the car to make sure Rosie was there and Mandi would not need a ride home. "Bye," Mandi called. "Thanks. Don't work too hard." "Not me," her friend said, laughing her cigarette laugh that turned into a cough, and then added before driving off: "Be good."

Rosie and Mandi watched television until the sun went down and then some. They danced to Mandi's favorite rap CDs, getting pumped for the night. Around eleven-thirty, Rosie and Mandi walked the few blocks into town so Rosie, recently promoted to gas station manager, could check on the girl working the night shift. On Highway 62, Rosie spotted a Marine in a white Ford Escort and waved him down. "Hi, ladies," he said, "what can I do for you?" "Oh," Rosie said, "I thought you were Majohn." "Nah, the name's Trent. Where you going?" "Just up the street," Mandi replied. He drove them to the Texaco and then back to Rosie's. "We're having a card party," Rosie said. "Wanna come?" Draper joined them. The three played a few rounds, then walked back into town to stock up on food and drinks.

The heart of Twentynine Palms beats in its fast-food parlors—connections are made, parties discussed, last words exchanged. Around midnight, while shopping at the mini-mart, Mandi ran into her mother and her mother's boyfriend. She gave each of them a big hug and then thanked Mike for making her mother happy. Mike liked hearing this; although they had been together for five years, Debie was not an easy read, and he knew that if Mandi said Debie was happy, it was true. Debie asked if Mandi had decided on birthday plans. Mandi did not want to talk about it, and Debie apologized. To smooth the moment, Mike gave Mandi some money for cigarettes. "I'll call you in the morning," Mandi said. "Okay," Debie said. "I love you." "Me, too," Mandi said. She and Rosie paid for their three bags of groceries and disappeared into the bright floodlights of the 7-Eleven parking lot. "Call me in the morning," Debie repeated, her voice swallowed by the night and the exhale of trucks and gears downshifting at the stoplight outside and all the other Mojave sounds, especially the sound of the desert itself, watching and listening and breathing.

At Rosie's, Trent laid down his cards and said he had gin, three of a kind and a run of four. He figured it was his lucky night, took a big, confident stretch, and sipped from a twelve-ounce can of Olde English. "My hand sucks," Mandi said, showing her cards—a jumble of numbers, except for a pair of twos. Rosie asked Mandi what was bothering her; she never did this poorly at cards. "Well, let's see your hand," Mandi replied, "if you think mine is so bad." Rosie displayed her cards. Two pairs—better than Mandi's but not memorable. "I think this Boone's Farm is fucking me up," Rosie said, taking another swallow of apple wine, "fucking me up real good. Trent, your winning streak has just come to an end." She picked up the cards and shuffled. Mandi swigged from her can of Dr Pepper and wrote down the latest score on a pad. Underneath the score she drew a heart and penciled in the name "Kevin," then scribbled over it. "Don't worry, Mandi," Rosie said, "there's plenty of fish in the sea." A couple of Marines showed up at the door, peering in through the screen, and calling singsong style, "Knock, knock." Rosie told them to come on in, and then to Mandi added, "What'd I tell ya?"

The taller one was Valentine, and he reminded Mandi that he was at Rosie's July Fourth party. "Oh, yeah," Mandi said, "you were hassling me." "That wasn't no hassle," Underwood said. "I was just being friendly." "Yeah, right," Mandi said. "Got any beer?" he asked. Rosie told him to look in the fridge and he grabbed a pair, handing one to his friend. They popped the cans open and sat down at the card table. "Five-card gin okay with you?" Underwood said to the group. Rosie shrugged why not and started to deal. "Hi, I'm Trent," the little guy said to Underwood and his buddy, extending his hand. "I'm Val, and this is

Carl," Underwood said, then looked Mandi up and down. "Hey, Trent, she with you?" "I can speak for myself," Mandi said. "I'm not with anyone." "Good," Underwood said. "That means you'll be with me soon." Mandi slapped her hand facedown and went to Rosie's bedroom in the back and dialed the phone. After listening to eight or nine rings, she decided everybody at the Crip house was smoked out. She gave it one more ring. Someone picked up. "Hey, Maurice," she said. "It's me, M&M." "Hey, baby, what's up?" he replied. She asked if he could come and pick her up at Rosie's. "This Marine named Val is trying to get stupid with me," she said. Maurice said that he wished he could, but his car had a flat tire. "If you hear from Bone," Mandi said, "tell him I love him."

Mandi returned to the card table. Underwood and his buddy were gone, but other desert predators were wandering the moonlit trails of Twentynine Palms. They began to arrive in small packs. A few more Marines stopped by, grabbed some beers, and cranked up some Coolio on the box. The Samoans from up the street made a customary noisy entrance—"Hey, where's the dancing girls?"—sat down at the card table, and Mandi added their names to the score sheet: Tony, John B, and Kaipo.

They played a couple of hands, and lost. Mandi penciled in another score for Trent. "You the man," Rosie said as Trent polished off another twelve-ouncer of Olde English. Tony pushed back from the table and opened the fridge door. "Rosie," he said, "I'm hungry. What you got here? I don't see nothing." Rosie told him there was a pound of beef on the top shelf. He took the meat and left. The other Samoans put their cards on the table and followed. Mandi left a few minutes later, trailing them to their apartment, a three-bedroom crash pad lived in by anyone who spent the night. "Hey, homes," they said when she walked in through the open front door. "What's up?"

Mandi joined them in the kitchen and said that she was sick

of Marines. "They're all such assholes," she said. Someone passed her a big fat joint. She declined, reminding the Samoans that she never smoked pot. Tony apologized, said that he thought that tonight might be the night that she really needed it. "I need to get outta here is what I need," Mandi said. Tony dragged off the reefer and ripped the wrapper off the beef, dumping the mound into a pan of popping oil and slathering it with packets of restaurant salt. "You guys are too fucked up to do this," Mandi said, and took over. She turned the gas flame down until the oil stopped spattering and scraped away some of the salt. She stirred the beef as it browned, chopped up the dregs of an onion, and mixed it in. She searched for seasoning in the empty cupboards and found a solitary bottle of ketchup with a sticky cap. She pounded the stuff out and let the desert dinner bubble and brown. "Smells good, homes," Tony said, grabbing a fork and scooping some up. "Yeah, you did good." The others gathered round the meat, and as they all chowed down, Mandi thought about how easy it was to please men, all you had to do was give them something—food or sex—and then they would appreciate you, make you feel wanted. She hoped Rosie was doing all right with all those guys in her apartment; even for Rosie, who liked having company, who liked being liked, being alone with a bunch of Marines might be too much to handle. She straightened up the kitchen and wiped the counters with a paper towel. "Thanks, Manders," Tony said, winking as she left. Mandi knew that when Tony used his special nickname for her, he wanted to spend the night together.

Blame for the grisly double homicide was immediately assigned to the usual suspects. Police mounted a sweep of the town's most unwanted, and fanned out across the desert. They questioned the Samoans who were friends of Mandi and Rosie, and, in fact, Tony Teo had aroused their suspicion during a polygraph test. "Are you withholding anything about what happened to the girls?" he was asked, and his inconclusive response triggered more questions. Yes, he admitted, he had been withholding information; he and Mandi had been making out and he had given Mandi the hickeys on her neck that the coroner had noted in his report. "I didn't want my girlfriend to find out," he told police. Others who were questioned included local Crips and Bloods, known drug dealers and users, anyone who had a habit of being out and about after ten in the evening, the town's curfew hour, and certain Marines, some of whom had dual affiliations—they were also gang members. Rumors characteristic of their bar of origin occupied the town. At Del Rey's, a shack with nice neon and a gnarly clientele on the outskirts of Twentynine Palms, everyone said, "It was a drug deal gone bad." The word was the same at the Josh Lounge and Al's and a few of the other dive bars, involving "shit happens" theories which pieced together Debie's past history with bikers who were recently released from jail, the rivalry between the Crips and Bloods, Marines running amok, and variations of all of these theories—each one resulting in the need to get rid of Mandi and Rosie as a warning to others. All of the rumors were believable and they confirmed one thing: the children of the desert were at great risk, easily caught in an endless Mojave cross fire.

When Mandi returned, Trent was alone at the card table, playing solitaire. Rosie lay on the shag rug in the living room, in tears, on the phone. "Where's Tim?" she cried. "Can't you get Tim?" "What's up?" Mandi said to Trent. He explained that Rosie had been trying to reach her boyfriend, "but he's out on the rifle range." "She knew that," Mandi said. "Well," Trent said, "she's been saying the same thing over and over." "Rosie,"

Mandi said, sitting down on the carpet next to Rosie, "come on, give me the phone." Rosie held tightly to the phone, kept crying. "Tell him I love him, okay? Please tell him that for me." She hung up the phone and then moved to the couch, wrapping herself inside her daughter's Mickey Mouse blanket, curling up baby-style and covering her head with a pillow. "Is she okay?" Trent said. "Oh, yeah," Mandi said, checking a wall clock. "Only a couple of hours until the sun comes up." "How 'bout you?" Trent said. "I'm okay. I guess." "Come here, maybe I can fix it." Yeah, right, Mandi thought as she dumped a mound of cigarette butts in a Raiders ashtray into a wastebasket. Tomorrow I'm gonna turn sixteen and you're gonna fix it. Tomorrow I'm gonna be sweet sixteen and then that's supposed to make me suddenly feel good, no one ever says sour sixteen or stupid sixteen, how come it's always sweet sixteen? What's so fucking sweet about it? Does it mean that suddenly I taste good and everybody's supposed to have some? What about last year and the year before that? Wasn't I sweet then? That's what everybody always tells me. "Mandi, you're a sweet kid." "Mandi, stay as sweet as you are." Yeah, right. Hey, you dumb sweet asshole Marine. Do me. Or let me do you. Who cares? Anyone who thinks they can fix this life doesn't have a clue.

The state called Trenton Draper. Debie turned and looked as he entered. He did not make eye contact. Debie did not know that Draper had initially been a suspect in the murders and may have at one time been offered immunity from a charge of the statutory rape of Mandi on the night of the murders in exchange for his testimony against the defendant. But no charges were filed and he was testifying not under a deal but through subpoena. He was sworn in and asked about his current employment. "I'm a chemical operator at Allied Chemical," he said, settling back a bit into the witness box, perhaps slightly relieved that there would be yet one or two more seconds in his life in which he did not have to describe in public what happened on the night he went to Rosalie Ortega's apartment with Rosie and Mandi, did not have to say in front of Mandi's mother, whom he had just befriended at the Best Western, what had occurred between him and Mandi. But eventually, the moment arrived. After the partying had come to an end, Draper said, he lay down on the bed next to Rosie. Then Mandi came over and began to fondle him.

Debie turned to Jesse Fulbright, who was sitting next to her, and flashed him a look, digging her fingers hard into her chair, hearing this news for the first time, wondering how Draper had been able to look her in the eye during their conversation at the Jolly Roger. Fulbright nodded in sympathy, and weariness; he had heard stories like this many times, and it always upset the victims, even those who knew it was coming, for of course the one person who could tell her side of the story was not here to tell it. "Mandi unzipped my pants," Draper told the jury. Debie started taking big breaths, and stared hard at Draper. When he finished his account of the evening, he looked neither to the left nor to the right as he left the courtroom quickly. Later at the Best Western, Debie tried to find him. She had heard that his plane wasn't leaving until early the next morning, and as she had put it, "It was *hasta la vista* for that little cocksucking motherfucker." But Draper had checked out ahead of schedule and gone right to the airport.

"Come on, baby," Trent said, "let me give you a massage." He finished off his Olde English as Mandi shrugged all right. Then she settled into his lap, facing him. They started to kiss, and he pressed his fingers into her shoulders. "You need this," he said. "I know," Mandi said. "Damn, you kiss good," he said softly. "You do, too," she said. "How old are you?" he said. "Old enough," Mandi replied. Trent urged Mandi down toward his hard-on and unzipped his pants. She sucked him until he was ready to come, then wriggled out of her panties and pulled him to the floor with her, just a couple of feet away from the couch where Rosie had rocked herself to safety. "Is this good?" she asked as he entered. "Do you like me?" "Oh, yeah," Trent said. "I like you, baby. You're a good girl." Mandi held him tight with her legs and arms, anxious for the quick ride out of town, and then he came, got up for another can of beer, sat down at the card table, and swilled it down. "Hey, Marine," Mandi said. "Come back and finish the job." "Can't, baby," he said. "I'm done." "That's not very nice," Mandi said. "Come on, you'll like it." He walked to her and lay back down. Mandi sucked on his dick again until he got hard and then she lowered herself down. "Yeah, baby," Trent said, but then lost his erection. He pushed Mandi off and rolled away. "Jesus Christ," he said. "You makin' me sick. I got a stomachache." He wouldn't look at Mandi, just stared at the wall, the one where Rosie had tacked up some pictures of Cindy Crawford modeling Calvin Klein in a recent *Vogue*. "Do you want something?" Mandi said, putting her clothes back on. "Water?" "No," Trent said, "leave me alone." He started to get up, then began retching until he heaved up a six-pack of Olde English and a night's worth of Cheddar-cheese sticks all over the carpet. Rosie stirred, changed sides, but did not wake. "Jeez," Mandi said, grabbing some paper towels from a counter and wiping up the mess. "See ya." After Mandi left for the Samoans' apartment, Trent dialed the phone. "Majohn," he said, just before stumbling to a mattress and passing out. "I'm fucked up."

"Trent told me he was at Fishhead's house," Majohn King recalled during testimony in Victorville, "drinking and playing cards. Other guys were there and there was a white girl." Draper and King were best friends in the Marines, having both been sent to Kuwait in the 2/7 and served on the front. Now working as a tool-and-dye setter in Oceanside and studying to become a cop, King explained that Draper had called that night because he did not want to risk a DUI with his friend's car, and promised that he would bring it back later. King himself was one of Rosie's admirers. "She had a great personality," he testified. "I visited her lots of times at the gas station . . . If Rosie and Tim ever broke up, we'd talk about getting together." Rosie got on the phone when Trent called, King testified, and said "they were over there having fun, nothing much was going on." King said he'd see his friend when he sobered up, and reminded Trent that they had plans to go to Orange County for the weekend. "When Trent got back from Rosie's," King said, "he was excited that he had passed out from drinking. He doesn't drink—we don't go to bars—and [this was a new experience]." Later that day, King continued, the pair headed down the hill, for forty-eight hours of freedom. "When Trent got the call in OC about the girls," King said, "he was like, 'Man get outta here. Don't be playin' like this.' He seemed like he was in disbelief."

August 1, 1991, was drink specials night in Twentynine Palms. As usual, shots of vodka and tequila were going for a buck and anything mixed was just twice that. Short of the river Jordan flowing with Budweiser, this was Marine heaven; if a boot couldn't get laid tonight, then maybe he should join the Navy with the rest of the girls. Lance Corporal Valentine Underwood had no trouble getting permission to leave the 3/11 barracks and go into town that night, although, as Tammy Watson believed, he was restricted to the base pending the outcome of the rape investigation, and had accordingly, as several low-ranking NCOs would recall in bars around town, been assigned "chasers"—Marines who keep tabs on other Marines. Underwood was a power forward on the Twentynine Palms Marine Base basketball team, regarded as the best player on the team, scoring an average of forty-five points per game. On August 2, there was an important game scheduled with the team's biggest rival, Norton Air Force Base. Underwood left his room and headed off the base and about twenty miles west on Highway 62, to the comparatively bustling town of Yucca Valley, home to Wal-Mart and JCPenney, the Morongo Basin's biggest stores, and several miles of strip malls whose sameness mimicked the surrounding desert. He turned into a block of stores on the north side of the two-lane, just past the blackjack place and the Jelly Donut. He parked his El Dorado and entered Jack's, a sports memorabilia store. Inside, beind a dusty glass case of trading cards, the owner happily greeted longtime customer Underwood, giving him a long, intense two-handed shake: "Hey, where you been? Haven't seen you, whaddya got for me?" Underwood said that he had something special, reaching inside his wallet and removing a card. "Check this out," he said. "Carlos Baerga. A rookie card." It was part of a special and limited series of cards that Topps had printed and shipped to soldiers stationed in Saudi Arabia and Kuwait during Desert Storm. "A stewardess from Kansas gave it to me," Underwood said. There was a pretty active trade in these cards once everyone got home, and someone like Baerga, a hot first-year player for the Cleveland Indians, could fetch about sixty dollars. "Oh, man," said the owner. "I'm sorry, but I got all the Baergas I can handle. I'm sorry, really, man, don't take me off your list . . ." Underwood asked if he would take it on consignment. "Sorry, man, can't do that." Oh, well, Underwood shrugged: "Snooze, you lose."

Underwood cruised back toward downtown Two Nine, disappointed that he wasn't able to pick up some extra cash for the evening. After stopping at Rosie's to recon the action, he made directly for the Club Max, on Highway 62 and Adobe Road. On drink specials night, arriving before ten meant no cover charge; the line to get in often snaked out through the parking lot and spilled into the street. On other nights, the place was often empty; in fact, the big dark box seemed haunted—juiced only by the blinking of video games, a swirling disco ball, and screens that flickered mass-market sporting events. Occasionally an urgent television voice echoed: "He shoots!" "He scores!" "And the Hail Mary pass . . . f-a-a-a-ils . . ."—the soundtrack to just about any of life's episodes, except here, as in all Mojave pit stops, the attempts to connect are just a bit more desperate, a bit more pressing, a bit more amped. Underwood figured he'd for sure get lucky on August 1 at the Club Max, and he was dressed summer rooster style—flowered rayon shirt, shorts, Italian loafers—open for business, along with everyone else, wanting to make an immediate exchange, expecting it would happen. To think otherwise, in fact, even to think about the odds at all, would have been as goofy as Saddam thinking he could win the war; the girls were not concerned with middle-class rules such as "wait until the third date before going all the way" and the boys weren't here to look for dates. In the desert, things just happen, biology takes over, and the vibes of such behavior can be felt in the winds when they blow in with nothing to stop them, can be heard in the life-sucking whine and fade of a distant freight train, can be seen in the aftermath of the quick drug buy in the parking lot at Del Taco, the dealer tearing off in an old Pontiac with the floodlight catching a front tooth capped with gold. The Club Max at the corner of Highway 62 and Adobe Road, behind a tattoo parlor and across the street from the Chemehuevi Indian cemetery with its unmarked graves, was the on-ramp to the thing that was about to just happen, the thing that the desert had been conjuring and contriving and warning about with its seismic turmoil and its far-flung keen which called to the most troubled in the land; the Max marked the fork in the road, it was right there at the split in the V, half a breath away from the house of horrors. The ancient storytellers in Twentynine Palms say that the Club Max is in fact haunted. It sits atop a grisly crime scene from another decade—in 1965, the tale goes, three satanic ritual murders were carried out in what is now the club's basement—and all who dance here dance forever on graves. This story is not known to most regulars, but would it matter if they knew it? People who hang here have always known they were out of the ball game; to them, it would figure that they danced with the dead.

"Hi, Phyllis, how you doin'?" Underwood said when he walked in that August night. He was talking to the big Samoan at the door who greeted all the customers and had come to be quite fond of the strapping Marine, a regular at the Max. Sometimes, Underwood would leave for a while, go to a Marine hang up the street called Stumps or across Highway 62 to a dive called Al's, then come back with a pizza from Rocky's for Phyllis. He was the only Marine who always asked her if she needed anything when he left for a round of barhopping. "Hey, homey," Phyllis said, smiling, "what's up?" And so began Valentine Underwood's last night on the town.

Once inside, he soon was chatting up two girls: Julie Schutte, a young mother of four living with a Marine, and Deborah Ashworth, a scrawny desert rat whose history was nothing but abuse—by acquaintances, by boyfriends. Phyllis knew they were party girls, small-fry pilot fish around a local speed and crack dealer who was a heartbreak to her father, a popular maintenance man. Underwood did what Marines all over the high desert were doing, since today was payday; he unfolded a wad of crisp new bills and asked the girls to name their poison. "Seven-and-Seven," said Julie. "Me, too," said Deborah.

"Make that three," he said to the waitress. The drinks arrived and he threw Phyllis a wink. It was going to be one of those nights, she thought, and she smiled back as in "Marines will be Marines." Many rounds and several close dances later, Underwood made his next move. "You know," he said to Deborah, "you remind me of my mother. Both of you are short." Deborah liked this; it meant that she was making him feel comfortable, pleasing him. They danced again, to something by Barry White. It felt good, right. A woman with long spiky hair and gnarly hands dressed in a hospital shift stepped out of the shadows. She was as old as the Mojave, the shade of one of the corpses said to lurk in the basement. "A rose for your lady?" she asked Underwood, unveiling a day-old bouquet she had swiped from a Kmart. Underwood looked at Deborah and she blushed. Yes, she wanted a rose. Her pleading eyes said that she wanted a rose very badly. Probably no one had ever bought her a flower before, let alone a rose. Underwood gave the peddler a five-dollar bill, took the rose, and handed it to Deborah. Silently crying with joy, knowing in her bones that no one would ever give her a rose again, she invited her Marine suitor to come home with her. The moon outside struck its 2:00 A.M. position: closing time. The desert hag moved off to ply her trade, under the swirling and glittering disco ball, back into the shadows of Twentynine Palms. Maybe the purchase of a rose would help a lonely Marine, far away from home and wondering what the hell he was doing in the middle of the Mojave, talk a love-starved girl into his pickup truck on this hot and empty summer evening.

Deborah, Valentine, Julie, and Julie's boyfriend left together. Their first stop was the 7-Eleven across the street, where Underwood talked the clerk into selling him a twelve-pack of Miller Draft even though it was too late for such an exchange. Then they headed for Julie's house, where Deborah was staying. The house was a short distance to the east of Twentynine Palms in a little berg called Wonder Valley. Unlike Twentynine Palms, Wonder Valley is too small to have its own newspaper, but has its own column in *The Desert Trail,* the Twentynine Palms weekly. It features local gossip ("Flo Connor—mother of Stinkey the Cat—entertained her son Butch at her home here . . ."; "We will soon have a Denny's restaurant at the site of the old Rings Restaurant at Four Corners . . . I am a great fan of their delicious soup . . .") and news of grange meetings and fire-department events. Residents whose comings and goings are reported only in police ledgers refer to the area as "Wonderful Valley."

Inside the small, failing pink bungalow in the middle of the Mojave, the kids were sleeping. Julie and John said good night to Deborah and her suitor and went to bed. Deborah headed for the kitchen, removed a six-pack of Bud from the refrigerator, and put it on the table, adding her own offering to Underwood's earlier purchase. She asked Underwood if he wanted to play quarters for beer. After he lost the second round, he demanded a kiss. Deborah declined, stating that she was not in the mood. He turned her head toward him and tried again. She backed away into a counter. Underwood reached behind her and grabbed a butter knife, wedging it against her throat. "I know why you asked me here," he said. He applied more pressure with the knife. Deborah pushed him off and ran to the bedroom door, banging on it. "Get this fucking nigger out of here," she yelled. "Get him out!" Underwood came after her and told her to shut the fuck up. She screamed. Inside the bedroom with their mother, the three children started crying. Julie opened the door in a nightshirt and emerged. Five-year-old Jasmine peeked out from behind her mother, her flat and weary brown eyes masked by the dark. "He tried to kill me," Deborah sobbed as Underwood stood behind shaking his head that it wasn't true. The other kids were crying, standing in a little row behind Jasmine. "Jasmine," Julie said, "tell them to be quiet and stay here. Let's not wake him up; he needs his sleep. Mommy will be right back."

On a November morning in 1997, Deborah Ashworth waited in the hallway outside Department D. She was a skinny wraith with stringy hair, minus various teeth. She had a black eye. She fidgeted and paced, waiting to be called as a witness for the prosecution. Her mother, squat with striking, long, perfectly white hair and a deeply lined face, the kind of skin that indicated years of desert living unshielded from the sun by creams or even a hat, paced and fidgeted with her. Mother and daughter had flown in together from Yuma, where they both lived in space 54 at the Whispering Sands Trailer Park. Deborah was nervous because she was wanted under an outstanding bench warrant elsewhere for a DUI and was afraid that by answering her subpoena to testify in Victorville, she would be busted. She was assured by authorities that she would be in and out of town quickly. She was also in fear for her life, dreading having to face Valentine Underwood and recount an incident that could help put him in jail for the rest of his life, and mark her for a revenge killing somewhere, somehow. What if he got out on parole? she wondered. What if he tracked her down? Her mother told her to just go in there and do the right thing and everything would be fine. The bailiff opened the door to Department D and called her name. She entered with her head down, followed by her mother. Her mother took an aisle seat, as it turned out, two seats away from Debie; Deborah proceeded to the stand and was sworn in. She began to recount her story. Her mother welled up. Noting this, Debie reached across the seats and held the woman's hand, although the two had never met.

When Deborah got to the part about Underwood threatening her with a knife, Hardy objected and the jury was dismissed for a hearing. "Your Honor," he said, "it should not be allowed under 352 because it is so prejudicial as to become a major factor in the jury's decision-making process." Bailey countered that under 1101B, Ashworth's testimony revealed the defendant's MO and intent. "A rejection leads to Valentine Underwood pulling a knife," he said. "When the defendant is rejected sexually, he undergoes a change. The probative value clearly outweighs the prejudicial, Your Honor." Judge Yent ruled in favor of the prosecution. The jury returned and the line of questioning continued. "I felt that Valentine Underwood was angry because I wouldn't kiss him," Ashworth said. Bailey asked her to demonstrate how the defendant had threatened her, by positioning her hand against his own throat. Underwood conferred intently with Lee Johnson as Deborah angled her hand against the district attorney's jugular vein. When she finished her testimony, Deborah and her mother quickly exited the courtroom. Debie followed them, trailed herself by Jesse Fulbright, and embraced Deborah in the parking lot outside the courthouse where witnesses and lawyers and spectators congregated and smoked cigarettes, or paced, or stared at a lone rosebush growing through the barbed wire that surrounded the holding pen where those on trial for various felonies were detained. "Thank you so much for what you did," Debie said to Deborah, giving her a hug. "You're my hero."

There was a wind advisory in effect in the high desert and tremendous blasts of hot, dusty air transporting the chemical particulates of vehicle emissions and the dregs of electricity factories in nearby Las Vegas and even further north from Salt Lake City swirled through Victorville, sucked up heat on the flats, and then picked up speed as they rolled down through the mountain passes, bristling the fur on coyotes and the feathers of ravens, all the way to the lower elevations where someone once gave them the name of Santa Anas; they deranged the chimes on the patios of Los Angeles and made the bougainvillea crackle and then, finally, at sea level, vanished with a flourish, carving nice hollow tubes in the breaking waves of the Pacific to many a surfer's delight, linking the desert and ocean in eternal perfection. A tumbleweed bounced toward Debie; once they were all over the West, now tumbleweeds were rarely seen, their disappearance another sign of something disturbing that scientists could not really explain. Debie kicked at the tumbleweed and watched it skitter off. Jesse Fulbright checked his watch and said that it was time to leave. He advised Debie to go back to the Best Western and get some rest. She declined and headed back to Department D. Deborah and her mother followed Jesse to his car, glad the ordeal had come to an end. He immediately drove them to the Ontario Airport to catch the next flight for Yuma, and they were gone before the jurisdiction which sought Deborah could find out that she had set foot in California.

A Gypsy once described his fellows as "crabs in a bucket." So, too, the nomads of Twentynine Palms, doomed to run into each other, to tear each other down for the rest of their lives, to bear each other's company, to entangle in each other's lives until they tired of the journey. The interchange between these few who had crossed paths at the Club Max did not end with the violent attempt to extract physical contact. Julie sat with Underwood in the living room and talked to him until he calmed down. "I was just playing," he told her. "I didn't mean nothing by it." Deborah returned to the kitchen table, finishing off another beer. Julie went back to bed. Underwood dozed off the booze. Deborah lay down on the couch and slept, too. At 5:30 A.M., Julie's boyfriend awoke to his clock-radio alarm, dressed, and headed for duty on the base. "I gotta check in for formation myself," Underwood said. John offered him a return ride. Deborah asked if she could join them. "I'd like to get out of the house for a while," she explained. At 6:00 A.M., the three passed the entry gate on base and Deborah accompanied Underwood to his room. Underwood said that as soon as he was through with formation, he would take her back home. But after she sat and waited for twenty minutes, the brief episode of solitude was too much to endure, and she needed to move, to be in motion, go somewhere, escape her past, which all seemed like the episode that just played out in Wonder Valley, and run faster than her future. She left the barracks. At the gate, she stuck out her thumb. "Where to?" said a young Marine in a new Thunderbird. "Town," Deborah said. He dropped her off at the 7-Eleven and gave her a few bucks for food. She bought some breakfast rolls and a giant-size Gatorade, her favorite hangover remedy, and returned to the street. She quickly hitched another ride. Back at Julie's, she swilled down the official drink of the NFL, pumping herself for the next act in life's dead-end play. The phone rang. She passed the jug to Julie, and answered. It was Underwood. "How the fuck you get way back out there to Booneyville?" he said. She explained that she got a ride and he said that he wanted to continue his visit. Maybe he wanted to apologize and start over, Deborah thought, gazing at the rose he had given her the night before; she had arranged it just so in the empty beer can on the kitchen table. Maybe this was the guy who would finally make things right. Okay, so he was an asshole. But even assholes can change. Everybody deserves a second chance. Right? She gave him directions.

Half an hour later, Underwood and another Marine knocked on the door and came in. "Look what I got," Underwood said, brandishing a bottle of cheap rum. He opened it up and they all took a swig. "Hey, man," Julie said, pointing to her three-year-old son, "it's my kid's birthday. I'm taking him to Chuck E. Cheese for dinner." "That's it?" Underwood said. "Can't afford nothing else," Julie said. "Man," Underwood said, "I know how that is." He handed his friend a crisp one-hundred-dollar bill and asked him to go get a six-pack and some toys for Julie's children. "You don't owe me nothin'," Underwood said to Julie. She appreciated the thought, she explained, because she wasn't that kind of girl. But Underwood persisted in his approach, suggesting that Deborah accompany his buddy to the store.

Deborah's heart sank—another trashed fantasy: Underwood didn't want her now that she had rejected him. She shouldn't have refused him earlier; now it was too late. Might as well leave with his friend, she thought, and followed him into his red de Ville. They headed out of the driveway, and about a mile away stopped on a sandy berm. The Marine broke out a crack pipe and lit up. "I don't believe in that shit," Deborah said. "Val said you was freaky," the Marine replied. "Just take me home," Deborah asked. "Jeez," he said, "you white bitches is all the same." Deborah started crying. "Now what?" the Marine said. "Nothing," she said. He stopped the car at the edge of the driveway. "Okay," he said, "party's over. Get out." Deborah walked up the driveway. Underwood was in front of the house, sitting on the skeleton of a sofa—a couple of sun-bleached, threadbare cushions atop a rusted-out mass of springs and wiring. He was smoking crack and didn't notice her as she walked right by. A few minutes later, he perked up—something was making noise—oh, it was his buddy in the de Ville, down there, on the street; time to go. He shambled toward the car. It was not the walk of a contented man, it was rangy, edgy, lurching: although drink specials night had come and gone, Underwood had not gotten lucky. Deborah watched the vehicle disappear as it headed away from Wonder Valley. Inside, she moved the beer can with the rose in it to a windowsill on the sunny side of the house.

A murder trial naturally carries the stench of death. In this trial, each criminalist dressed as if he or she were death's very messenger, each bringing the grim tidings not just with words but with the very flesh itself. In addition to David Stockwell, the DNA authority with a distinctly protruding upper lip and slow walk, there had been Valerie Saleska, the prosecution's blood-typing expert, whose skin was almost totally devoid of color. Now came Karen Anne Rice, also pale, and dressed with a spooky, theatrical flair. She approached the stand with much less trepidation than some, almost with a swagger. She was sworn in by the bailiff. She threw off her black cape. Underneath it she wore a long, dark flared skirt, a frilly, white Victorian blouse with a high collar, and a royal-blue cardigan buttoned up over it. She noted that the room was chilly, which it was, although she was the first witness to say so for the record, and the only witness among those who frequented refrigerated evidence lockers to note the temperature.

Gary Bailey was also dressed with particular flair on this sharply lit fall morning. He wore a gray sharkskin suit and mauve shirt, clothing that would befit the introduction of a most startling piece of evidence. "What happened on August second, 1991?" Bailey asked. "My supervisor told me to respond to a call in Twentynine Palms," she said. "I was part of a team." She explained that the team gathered and cataloged evidence, going over the exterior and interior of Rosie's apartment for latent fingerprints and blood. "B6 on the evidence list," Bailey said. "What is that?" Rice said that it was a section of wall from Rosie's apartment. It bore the imprint of a bloody human palm. "You could see the ridge detail on the palm," she said. Bailey entered exhibit 38 as evidence, a photograph of the blood impression made by one of the criminalists. By now, the jury knew the routine, knew that the introduction of a photograph of a certain piece of evidence meant that the item itself would most likely be introduced next. They leaned in, anticipating the unveiling of the bloody handprint.

"Your Honor," Bailey said, checking the wall clock above the defendant's side of the courtroom, "it's almost lunchtime. I suggest that we break now and resume in one hour." Court was recessed and the various players proceeded to their usual lunch spots. Bailey went to his office in the courthouse building. Judge Yent retired to his chambers. Escorted by the bailiff, Underwood was taken into back quarters where defendants in ongoing trials gathered for lunch. The jurors had formed little circles of friends, as groups of people generally do. There was one group that generally lunched at the local Greek restaurant, which by consensus served a very respectable hummus. Another preferred to economize, and repaired to El Pollo Loco. There were one or two who generally brought bag lunches and, weather permitting, would sit outside on the lawn. Several often went to the coffee shop at the Best Western. So did Debie, Jessielyn, and occasional visiting witnesses. One juror in fact had parked his RV in the Best Western parking lot and was planning to live there for the duration of the trial. But he was asked by Judge Yent to live elsewhere, as his regular presence at the motel, where witnesses were living, made it difficult to maintain the traditional wall of separation between jurors and participants in a trial. So he decided to sleep at home every night, even though it meant a three-hour round-trip drive every day, from a hamlet without a name in the San Bernardino Mountains to Victorville. But the Best Western remained his preferred lunch spot. Generally, the jurors and witnesses followed instructions and remained without expression, or averted their eyes, when walking past Debie and Jessielyn at lunchtime or in the courthouse hallways. Today was no different, although everything felt different: the energy of anticipation about what the prosecution was about to unveil hung in the air-conditioned climate of the coffee shop, perhaps elevated the temperature a degree or two; in fact, the place seemed to crackle—eyes were more furtive than usual; people moved quickly. Ronnie, everyone's favorite waitress, had overheard one witness say to another that Bailey was going to show the bloody palm print after lunch. The jurors at the table near the window facing the main parking lot, which was five tables away from the table where the witnesses sat, knew that Ronnie had heard something interesting because they noticed an increased edge to her body language, saw her lean in just a little more closely than she usually did when listening in on witness conversations, stop writing in mid-order for a second or two. So did the hostess, who had struck up a friendship with the juror with the RV. It was a strange game of criminal justice telephone, in which no one could tell anyone what they had just heard, but their body language passed on information that something big was up. What would the palm print look like? everyone wondered. Would it drip blood? Some of them could not eat. Others wolfed down their food and hurried back to the courthouse a few minutes early.

All players returned to their seats. Karen Rice was called back to the stand. Bailey asked her to step down. He handed her a large envelope, about twenty by thirty inches, and a pair of scissors. "Would you open that, please?" he asked. She cut along the rim very slowly and carefully. She carefully slipped out the contents of the envelope. She cut away the property tag. "What is that?" Bailey said. Rice identified it as a piece of wall board. She placed it upright on the witness stand and resumed her seat. To the left was part of a poster of a young girl at the beach. To the right were the faint outlines of red splotches. The splotches had faded over time. "Can you describe that?" Bailey said. "It's a palm print," Rice said. ". . . I believed it was blood." As she continued her description of the palm print, unveiled here for the first time in six years, Rice was visibly and audibly overtaken by a malevolent force. Her face drained of what little color it had and she began to sound as if her nasal passages were closing, or filling with something that was not air; within minutes she resonated with the rheumy tones of a bad cold, or perhaps she actually had contracted one. As she continued to testify, one had to lean forward to hear her diminishing voice; she was weakening, just as those who were the first to enter King Tut's tomb gasped their last breath upon breaking and entering into the forbidden and breathing the fetid air of time.

Upon the conclusion of Rice's testimony, the section of wall, and the print, exhibits number 85 and 85A, were entered as evidence. The jurors were invited to file past the print and observe it closely as Rice held it aloft. Some paused to put on their reading glasses. One—the most distinguished-looking member of the jury, a lean, bearded, middle-aged black man who always wore a stylish suit—lingered for some time, stooping to examine the print, checking the angles. When they finished, Rice returned the bloody impression to its paper crypt, sealing up for eternity this evidence of murder most foul.

Too drunk to return to base, Underwood asked his friend to drop him off in front of Adobe Liquors at the pay phone. "I thought you said those girls was freaky," his friend said as Underwood stumbled out of the car. The pay phone at the corner of Adobe and Two Mile roads has conveyed its share of desert distress. It was here that Tammy Watson reported that she had been raped by Valentine Underwood, here that Underwood called 911 and instructed the operator to ignore his victim's message. On the morning of August 2, 1991, Underwood dialed a taxi company. The dispatcher often received calls late at night from Marines and, in this case, recognized the voice of the man who was calling from Adobe Liquors. "Hey, Val," the taxi dispatcher replied. "How you doin'? Need a ride?"

The cab would arrive in ten minutes, and Underwood sat

down on the gravel next to the phone booth and stared at the morning sky. It was clear, empty. A couple of jets from the base screamed overhead and danced with each other, then disappeared in the vapors. Now Underwood's eyes lowered, toward the highway—no cab in sight, just more airborne activity, a squad of ravens now, breaking formation to revel in the overflowing Dumpster in the parking lot. Perhaps one of them tore off a piece of—what?—a stale mini-mart sandwich, maybe, or a strip of beef jerky that had been cast off by a tweaking speed freak. Like the coyote, the raven had learned to flourish in the desert, feasting on the lefttover preferred nourishment of the humans who lived in it and, in fact, increasing so much in number because of the proliferation of garbage across the desert that it was now a threat to the desert tortoise, whose hatchlings it would scarf on the way in or out of town.

A young black couple with dreadlocks pulled up in an old VW van painted with psychedelic flowers. The man got out and headed for the store. Underwood asked the man if he was Bob Marley. The man laughed and kept walking. While he was inside, Underwood approached the woman. "What's up?" he said through her open window. "Not much," she replied. "We're going into town to the cash machine. Nice morning, huh?" "Yeah," he said. "Quiet. I need some quiet." The man came out of the store with a few bags of groceries and offered Underwood a ride. Underwood got into the van, stepping behind the passenger seat and sitting on a shag carpet on the floor. "Know where I can get some coke?" he asked. The pair said that they didn't. "Let's go to Palm Springs, I know where I can get some," he said. The pair declined. At the bank, everyone withdrew some cash, and then Underwood asked the pair to stop at an apartment around the corner. "There's always a party going on," he said as he got out of the van. "You know how the Samoans love to party." He knocked on the door. No answer. He knocked again.

Finally someone answered. "Hey, man," the Samoan at the door said. "Party's over. Everybody's sleeping." Underwood barged in. "Got some weed?" he asked. "We don't have no more weed," the Samoan said. "Where the beers at?" he asked. "There ain't nothing here," the Samoan replied. Underwood wandered down the hallway of the apartment, opened a bedroom door, and turned the light on. There was Mandi in bed with Tony Teo. Underwood laughed and turned out the light. As he shut the door, Mandi told Tony that she was worried about Rosie. But Tony was sleeping. He didn't hear Mandi when she said that the big Marine named Val, the one who was trying to get stupid with her, was right here in the apartment.

Outside the apartment, Underwood shrugged nothing going on here at the couple in the van, thanked them for the ride, and walked off. Across the street, he checked the Bowladium for the dregs of any action or perhaps the start of something new. No one was there. Then the big, drunk Marine—alone in the bright and happy Mojave sun, rejected by the destitute, his violent attempts to connect achieving nothing, very possibly the only man in Twentynine Palms to have failed on drink specials night—headed down Palo Verde. Maybe something was happening at Rosie's. Yeah, he'd try Rosie's. There was always something going on at Rosie's. Always a party.

"Tony, I gotta go," Mandi said, kissing him on the cheek and climbing out of bed. "I gotta see what's up with Rosie." She got dressed, took three dollars from her wallet, and put the money on Tony's dresser. "For cigarettes," she said, and closed the bedroom door, stepping quickly over the sleeping bodies in the living room and heading for her friend's apartment down the street.

Aveola Teo, or Ave, as she was known to her friends and family, was Tony Teo's sister. At the time of the murders she was living with her brother. "The first time Mandi and I met, we beat the shit out of each other," she recalled in the hallway outside Department D. She could not remember what triggered the fight, but recalled the moment fondly; from that point on, she and Mandi were good friends. Ave was big, much bigger than Mandi, about six feet and easily 250 pounds. "Ave was good for Mandi," her brother said. "Ave's good people and so was Mandi. After the fight, they respected each other, you know how that goes." Ave was extremely nervous about testifying. Her face was wet with sweat and she fanned herself with a newspaper. Rick the bailiff opened the courtroom doors and motioned for her to come in. She sighed and turned to her husband, a thin man about her height. He blew into her long black hair to cool her off and told her she'd be okay.

On the witness stand, she became very agitated when she looked over at some of the exhibits and saw Mandi's ninth-grade class photo. "Toward the end I was in daily contact with Mandi," Ave recalled. She testified that she had been dozing off on the living-room floor when Mandi left for Rosie's early that morning. "I had been drinking," she said. "I was not drunk—I was buzzed on Budweiser. I don't think I had ingested any drugs. When Mandi came out and said she was leaving, I said I didn't want her to go back. She said earlier some guys were bothering her—Trent and Valentine—they touched her butt, they were both drunk. I told her to go back to Tony's room. She did. I went to lay back down. She didn't stay that long—I guess she sneaked out. I never saw her alive again."

There is nothing more promising than the desert in the morning; the chirping of the cactus wren and the scent of warming creosote hint of a life that is everlasting and right now. Mandi—back in her element, happy to be checking on her friend—got to

Rosie's first and walked in. As often happened, the door was unlocked.

"Hi, homes," Mandi called out. From the bathroom where she was brushing her teeth, Rosie called back: "Hi, homes. What's up?" In a few minutes she would be leaving for the morning shift at Texaco. "You know me," Rosie continued. "Us fish heads gotta get to work, otherwise this country would fall apart." Mandi walked past the open bathroom and told Rosie she had a bad feeling, wanted to check on things, and that's why she had come by. "No worries," Rosie said. "Tim's gonna come by later." Mandi was relieved that the day actually was starting out like every other summer day in Twentynine Palms, like every other day in every other town—plans were made, people went off in pursuit of the various ways in which they would secure dinner, nothing had gotten better, but last night was over and done with. "I'm gonna take a nap," Mandi said, and went into the bedroom to lie down. "I'll see ya later," Rosie said. "We'll figure out what to do for your birthday." She stepped into the shower, turned on the water, and closed the white curtain.

A few minutes later, Valentine Underwood opened the door and crossed the perimeter of Mandi's magic circle.

During the Gulf War, there was something that Underwood and some other black enlisted men in the 3/11 often discussed. As they saw it, there was a disproportionate number of blacks on the Kuwaiti front. There was a disproportionate number of blacks in their unit, the 3/11, one of the first Marine artillery units to participate in Desert Storm. Underwood had even mentioned it to his mother when he got back, pointing out that a number of black Marines had converted to Islam after the war, having decided that the religion of their Arab enemies was a better fit than whatever form of pro-American Christianity they had been practicing before joining the Corps. Zimena Underwood counseled her son to continue to have faith in the Bible—it could get anyone through the darkest of times, even the hardship of having the wrong color of skin in the land of the free, especially while living in the Mojave, which Valentine had told his mother was the most racist place he had ever seen. The white guys on the base were racist, he said, the cops in town were racist, the waitresses in the local restaurants wouldn't serve a black man, local shopkeepers followed black men around their stores lest they pocket the merchandise, the whole fucking wasteland was worse than the deep South because in the desert no one cares about anything, that's why they live here—he hadn't met one black person who had moved to Twentynine Palms voluntarily.

The knock on Valentine Underwood's door came early in the morning of August 3. He was asleep in his bunk aboard the Balboa Naval Base in San Diego, where he had been sent for treatment of his injured left hand. When he opened the door, two military investigators told him to pack his duffel bag, not replying when he asked, "What is this shit?" He followed them through the corridor of the barracks, down a flight of stairs, and into the lobby. The pair from the Naval Criminal Investigative Service turned him over to another pair of cops, two members of the San Bernardino County Sheriff's Department, Sergeant Tom Neely and Sergeant Jim Palacios. "Let's go," they said, and escorted Underwood to their car. "Hey, what is this shit?" Underwood asked. "This have something to do with that bitch say I raped her, 'cause I straightened that shit out."

He did not know that it was the thing that he loved, basketball, among other things, that had led police to him, the fact that like so many athletes, like so many men, he had an idol and that idol had a number and that number happened to be thirty-three and it was his worship of this idol who wore the number thirty-three that gave him away. For as it turned out, although various witnesses had placed Underwood at Rosalie Ortega's apartment on the night of the murders, and although the police knew he had a deep cut on his left hand before they were sent to pick him up, it was not until military investigator David Hertberg, a basketball player himself, was assigned to the case that an early break came. The coroner had tallied up the number of knife wounds on each victim. Police had pondered the significance of the number—thirty-three. Was it some satanic thing? Hertberg, lean, slight, and fast, was on the Twentynine Palms Park and Rec League team. So was Underwood. Hertberg was in awe of the Marine's athletic talent. He was also intrigued by a coincidence. Upon hearing that Underwood had been placed at the crime scene, and learning of the number of times that Mandi and Rosie had been knifed, Hertberg immediately recalled that Underwood's number was thirty-three. Underwood wore it on his shirt. He was proud of it. Maybe it intimidated people because it evoked Patrick Ewing. Hertberg faxed the information to his fellow investigators at Balboa. They were to hold Underwood until Neely and Palacios got there.

Detectives Tom Neely and Jim Palacios sat with Valentine Underwood in a makeshift interrogation room at the Balboa Naval Hospital in San Diego. The three formed a picture of the modern American melting pot—not the kind that liberals envision, with representatives of ethnic diversity living happily ever after in harmonious and picturesque neighborhoods, but the way it really was: a gringo, a Latino, and a black guy all hanging out together because there was trouble at the low end of the classes. California native Neely was lanky, good-looking, had blond hair, blue eyes, a mustache. Jim Palacios was short, beefy, fast, with slicked-back black hair. A Hollywood buddy movie might have called them *White Man and the Beaner*. And now, having survived, after a fashion, the fires and the toxins and the whole strange sojourn of death and destruction in Iraq that most Americans remember as a quick air war, Underwood—the big black Marine from the urban East Coast, locked up deep in Southern California where vast hordes of palefaces hunkered down in gated communities, in true enemy territory—began the street fight of his life.

Initially, Underwood had not been told why he was being taken into custody. The interview, like many police interviews, began with a kind of formal friendliness. "Okay," Neely said after queuing up the tape, "uh, first of all, get comfortable here, uh, what happened to your hand?"

"Well," Underwood replied, lighting up the last of his Kools and dropping the match to the floor because there was no ashtray, "I had whopped my hand the last two weeks, I been hurting my hands. I cut one there . . . playing basketball, up on the backboard where the pad's missing. I hurt this one over here . . ."

Neely asked him which hand he was talking about and Underwood said his right, adding that the skin had come off from playing basketball. "I hurt this one," he continued, indicating his left hand, "over here . . . I hurt [it] on the corner of a backboard and I also hurt it cleaning out the bucket on, ah, for field day and the razor blade." When pressed by Neely, Underwood explained that field day is "when we have to do a spotless cleaning of your room and everything . . . [it] basically consists of taking everything out and doing the corner of the room from top to bottom. Uh, you have to do GI cans, which, ah, means trash cans has to be scrubbed inside out, everything. I reached in there scrubbing it out, a razor blade cut me on this hand . . . This happened on Thursday, the razor blade cut me. Uh, cleaning that out and over here my left was, ah, real swollen up. That happened around eight days ago or something. Well, you know—no, about ten, ten or twelve days ago. And basketball. I been soaking it, I got permission and been soaking it for the last week and a half from, uh, from people back here on base . . . I been soaking and trying to get rid of it, trying to take care of it . . . I was basically trying to baby my hands, trying to get them healed for the tournament."

Again, Neely questioned Underwood about the time of the cut. Underwood reiterated that it was on Thursday, "about, ah, shoot . . . Thursday about . . . phew, about . . . about, about seven o'clock." Neely asked Underwood if he was referring to the morning or night. "Night," Underwood replied, letting the ashes fall from his cigarette to the floor. "My—my roommate, uh, we made a deal where where he do the, the room this time and I do the head. And, uh, when I went in to do the head, I had my, see our head is adjoined to two rooms so I had to do two people's rooms. Two people's cans 'cause they had they cans in there, too . . . I bitched at him about it, 'cause I think it was their can that had the fucking shit in there."

"What, uh, what kind of razor blade was it?" Neely asked.

"It was, uh, well it looked like it was the blue part of a razor that had peeled back," Underwood said. "It looked like where someone had tried to clean it."

"I'm—I'm not sure if I follow you."

"You know the, the razor, that's—"

"Like a shaving razor? Uh-huh."

"Yes, sir."

"The plastic ones?"

"Was the blue part," Underwood said, "plastic part."

"Like those Bics?"

"Yes, sir . . . The half, the top part of one of 'em. It wasn't the handle . . . It looked as if one of them had been messing with it. And I had, ah, that's what I asked him about. The fuck, it would, it looked like somebody had been trying to pry at the thing or something."

"Okay," Neely said. "Um, when did, when did it finally,

when did you finally end up getting medical attention for it?"

"Um, Saturday," Underwood said, just about done with his cigarette, "they looked at it Saturday. I was playing in the ball game and I kept playing, I was trying to play, 'cause Thursday, Thursday, uh, we didn't have to practice. We practiced on Wednesday and they had, uh, known that this hand was bothering me, pretty bad. My, ah, right hand was bothering me."

"Did anybody else know?"

"Excuse me?"

"Did anybody else know," Neely repeated, "that your hand was bothering you?"

"Yeah," Underwood said, "uh, the coach and them knew that my hand was bothering me. They seen me come out, 'cause this hand had the same, the skin off over here. And, uh, they looked at it and everything and he know that I hurt it, uh, Tuesday. And, uh, everybody was out there playing seen, ah, 'cause it kept bleeding. It kept, keeps bleeding over and over. And, uh, they seen it."

"Okay, so they knew that you had had problems with this hand," Neely said, referring to the right, "but they didn't know anything about this hand? They knew about your right hand but they didn't know about your left hand?"

"Well, they knew this hand," Underwood said, "but they didn't know, this wasn't cut then, this was just swollen . . . See, that's what I been saying, sir, the last few weeks I been bustin' my hands up. Been down bad," he said, then mumbled something about his knuckles and coughed. Neely offered Underwood his cup of water. "But," the detective said, "I gotta warn you. I've been sucking on the ice cube because I got a cold coming on."

Marines tend not to want to discuss bad news unless it can be fixed. For instance, the shadow side of the successful recruiting commercial for "the few, the proud" is the following statistic: between the years of 1980 and 1993, more members of the Corps died in accidents, homicides, and suicides than in any other branch of the armed services. "Whatever anyone else does," say rank-and-file Marines, "we do bigger and better." Another taboo topic is Gulf War syndrome. As far as the Corps is concerned, it doesn't even exist. To suggest otherwise in the Marines is worse than it would be in other branches of the military because it violates the code of Semper Fi. The acknowledgment also wreaks havoc in civilian culture by suggesting that the decisive American victory against Saddam Hussein came at a price greater than money. But among the 700,000 troops who served in the war in Kuwait, thousands have found their symptoms impossible to ignore and have reluctantly come forward with reports of memory loss, balance disturbances, sleep disorders, rashes, exhaustion, body pain, and chronic diarrhea. One study suggests that some of these symptoms are a result of brain damage incurred by exposure to nerve gas, the insecticide DEET (present on flea collars which soldiers wore to repel pests), or the drug pyridostigimine bromide (PB), which was given to at least 250,000 troops to guard against the toxic effects of nerve gas. Gulf War veterans with brain damage may have a genetic vulnerability to certain chemicals.

Former Lance Corporal Robert Windle did not look well on the day that he testified in *The People* v. *Underwood*. He was pale and thin, and walked slowly, appearing not particularly up to the task of testifying. Jesse Fulbright asked him if he was all right. "Yes, sir," he said. As he waited in the hallway outside Department D, he broke out in a sweat and excused himself. A few moments later, he mentioned to a visitor that his flight from Tulsa (he had to change planes three times) had left him feeling jet-lagged. Windle had been called because he shared an adjoining bathroom with Underwood on the base at Twentynine

Palms. On the witness stand, he said that Thursday, August 1, was a field day. "We started early, at around three in the afternoon," he recounted. "We filled the trash cans with Pine-Sol and water and mopped the floors. I put the water in the can that day. Prior to that I cleaned it out to make sure there was nothing there. The defendant did not come up to me and bitch about cutting his hand on the can. Yes, we shaved with blue Bic disposables, but I saw no razors or remnants inside the can on August first, prior to putting water in the can. The can was not rusted." Bailey asked Windle whether he saw the defendant the following day. "I saw him at eight A.M.," Windle recalled. "He asked to borrow money. I gave him ten dollars. He asked for twenty. He said he had to go to a basketball game . . . Valentine Underwood could come and go as he pleased." After a brief cross-examination, Jesse Fulbright escorted the witness out of the courtroom and drove him to the airport so he could go right back to Tulsa, where he worked in shipping at Sears. Although he had gotten time off to testify, it was just for one day. He had not wanted to use up one of his vacation days to stay overnight in Victorville because his wife was nine months, two and a half weeks pregnant, obviously due at any moment, and he wanted to use whatever extra time he had to be with her and his new baby, even if it meant taking three flights home instead of the two if he had waited until the following day to travel. And he was still not feeling well. "I have Gulf War syndrome," he told a visitor as he looked away, as if in shame. "I get the shakes."

Lance Corporal Valentine Underwood was about to mount a classic Marine raid, employing shock, surprise, and violence. His entry into Rosie's apartment was muffled under the drone of the swamp cooler. He got rid of what light and sound he could control; he shut the blinds and turned down the radio. He quietly unbuttoned his shirt, took it off, removed his shorts, slipped off his loafers, removed his socks, and put each one into a shoe. He laid his clothes across the back of a folding chair at the card table. He stepped quickly and silently into the kitchen, opened a drawer, and grabbed one of Rosie's cooking knives, the one with an eleven-inch blade, the one she used to chop up vegetables to make her lumpia, and headed past the bathroom and its closed door. He entered the bedroom. Mandi, still in her bikini top and shorts, was beginning to doze off. He stood over her for a moment and she felt a disturbance in the air. She opened her eyes. She started to bolt, but before she could get past him, he ripped an electrical cord from a lamp, tied her hands behind her back, and pushed her down on the floor, forcing her bikini top up around her neck and tearing off her panties, excited by the act. Then he straddled and raped her. While she was screaming, he plunged the knife deep into her neck and then sliced it across her throat, through the larynx almost to the spine. The first blood leaped out and attacked his chest. He was battling an enemy, trying to vanquish the resister—an ancient war cry pounded in his brain—"*War to the knife, knife to the hilt*"—not something that he had learned at Parris Island, but something that dwells in the heart of all killers, the thing that takes over when any man is roused to war. Mandi's feet flailed for a moment, her torso lurched, and then her body stopped reacting; with her struggle over, he kept cutting, at the pelvis, buttocks, mostly at the breasts, over and over again until he was spent.

The most graphic testimony began on the morning of October 22, 1997. Dr. Frank Patrick Sheridan, a forensic pathologist and chief medical examiner for San Bernardino County, began his presentation of the autopsy report. Lacking the on-air personality of the coroners who have become celebrities by way of frequent television appearances in recent years, Sheridan was nonetheless engaging, with his Irish accent, sad, deeply wrinkled face, and rumpled suit that hung heavy, seemed to carry the weight of a lifetime spent in the county morgue. Many in the jury leaned forward as Bailey unveiled a series of photographs. "Mandi is here and telling us what happened," he said, and turned to Sheridan to elicit the story of the wounds. Sheridan began, using language that was foreign to the modern Mojave, but not out of place in the desert a century ago when cowboys and all manner of itinerant free-rangers freely quoted Shakespeare. "Well, firstly," he said, he looked for the number of wounds, evidence of a struggle, what kind of weapon was used, and whether the wounds were consistent with the weapon. In the case of Amanda Lee Scott, he immediately noticed that the weapon was a knife and that some wounds were incised—pushed along the skin. Pointing to photograph number 70, he explained that this was a wound that cut the larynx. Seventy A depicted deep neck wounds in which major blood vessels were cut. "Mandi was initially able to continue making some sound," Sheridan said. Seventy B depicted wounds to the collarbone. "Here," Sheridan said, "the knife was stopped by bone . . . There is quite a lot of force associated with these wounds." Seventy-one A and B showed wounds to the central and left side of the chest, as well as the left armpit. Sheridan explained that there were wounds of different types on Mandi's chest; some were in the shape of a chevron, some were slit, others were linear. Some wounds went through the ribs and cartilage and there was damage to the heart, lungs, and diaphragm. The knife had been plunged through the chest cavity to the back, a full eight inches. One wound had been stopped by the sternum. "The bone brought the knife to a sudden stop," Sheridan said. There were wounds into the left and right lung, a wound to the spleen, upper abdomen, and through to the back of the liver. "A single stab wound to the lung," Sheridan said, "is sufficient to cause death." As he revealed the number of wounds—thirty-three—Underwood put his head down. It was time for exhibit 72, a bag bearing a tag (item of evidence number C33672) that said CAUTION KNIFE INSIDE. Bailey turned to his investigator for the day, Detective Norman Parent, and asked him to open the bag. Parent extracted exhibit 72A. It was the knife. He handed it to Bailey. Bailey showed it to Hardy. As Hardy examined it, Underwood ran his eyes across the weapon. Bailey handed the blade to the coroner. Sheridan held it aloft. Debie gasped. Some jurors were taken aback, eyes bulging at the sight of it. "Mandi would have been dead within a couple of minutes," Sheridan said, summing up the attack he had just described. Debie braced herself as a mannequin of Mandi's head and torso was now entered as evidence, placed atop a stand before the jury. It bore a mark for each wound, approximating where it was inflicted and what it looked like. The jury studied the mannequin. Bailey questioned Sheridan about other aspects of the autopsy report, including bruises around Mandi's vagina. Sheridan said that there was a fresh hemorrhage, consistent with sexual activity at or near the time of death. In other words, she had been raped.

War to the knife, knife to the hilt. The battle was not over; there was another enemy on the field. He got up, still holding the knife. As Mandi's blood oozed and flowed away, forming rivulets that ran everywhere, sinking down and down and down through the shag of the carpet, through the flimsy foundation of the bungalow, consumed as soon as it dripped onto the very sands which waited so patiently for quench, Underwood entered the bathroom to complete his mission. Rosie was still showering, rinsing her hair. He sliced through the white curtain and grabbed her, hauling her out onto the tiles, again excited, raping her from behind next to the toilet. *War to the knife, knife to the hilt.* As she fought for her life, he sliced furiously at her arms, then at her throat, most viciously at her breasts, just as he had done to Mandi, again and again and again. When it was over, she had been stabbed thirty-three times. Underwood stood over Rosie's bleeding body, then returned to the bedroom, carrying the knife, to take one last look at Mandi. An air bubble escaped from Mandi's open larynx and the sound, kind of a prolonged growl, scared Underwood and he jumped back, bracing himself against the wall adjacent to Mandi. What if she wasn't dead? he thought. Maybe he better check. He walked toward her and knelt, prepared to cut again if he had to. He placed his ear to her chest; heard nothing. He picked up her wrist and felt for a pulse. She was dead. He rose, carrying the knife, stood for a moment, and looked at his handiwork. Satisfied with both kills, he wiped the blade with a rag and dropped it. He walked back through the hallway, again stepping over Rosie and into the bathroom, where he quickly showered and toweled off. Coming into the light of the living room, once again past Rosie, he squinted as the morning sun began its blast through the cracks in the blinds. He got dressed. A Mickey Mouse clock in the kitchen said ten o'clock. Finally, as drink specials night had faded into day, the Marine had scored. He sat down at the card table and put on his Armani loafers.

Bailey moved on to the autopsy of Rosalie Ortega. Jessielyn Gonzalez was sitting in the courtroom. She sat next to Debie, who in turn was sitting next to Jesse Fulbright. Jesse told Jessielyn to take a deep breath and prepare for what was coming. "If you have to get up and leave, it's okay," he said. "It happens all the time." Bailey introduced Exhibits 75A and B, close-up photographs of breast, neck, and chest wounds. "There is movement indicated when the knife went into the breasts," Sheridan said. "The wounds are in different shapes, like they are on Amanda Scott." Sheridan reported that the knife hit Rosie's sternum. It also penetrated rib cartilage. The wounds to Rosie's neck were not as deep as those to Mandi's. But the abdomen wounds penetrated into the liver and its blood vessels. Photograph 75B portrayed Rosie's back, with wounds that had gone through her abdomen to the kidney.

The jury returned after a lunch break to another dramatic unveiling—Rosie's mannequin. It was right next to Mandi's, marked up with wounds. Jessielyn started weeping, just loud enough for the jury to hear. Debie tried to calm her; emotional outbursts, they had been warned by Fulbright, were not welcome in the courtroom and could hurt their case. All together, Sheridan reported, Rosie was stabbed thirty-three times, just like Mandi. "The wounds are remarkably similar," Sheridan said. Bailey began his linkage of the murders to Underwood. He introduced exhibit 78, the defendant's medical records, and exhibits 79A to C, photographs of cuts on Underwood's left hand. "The cut was very consistent with injuries inflicted on Mandi and Rosie," Sheridan said. "There are at least six wounds to the hand . . . The wound in the palm is quite deep. The median nerve in his palm was cut."

On cross-examination, Hardy questioned the number of wounds. He suggested that one wound could actually have been two or three. "For instance," he said, "Mandi's major neck wound was three to four inches in length. The knife appears to have been oriented both ways, looking at the streaks at either end of the photograph, indicating a repetitive movement, backward and forward. It could have been two or more slices, certainly more than one wound . . ." Sheridan disagreed, reiterating that he counted exactly thirty-three wounds on each victim. Asked if the killings were a result of a frenzied rage, he said that they were not; the stab wounds would have been more clustered together than they were in Mandi and Rosie. "This is a very violent attack," he explained, "with an intent to kill. I have some reservation about the word 'rage.' The attack was rapid but quite spread out. The attack was brutal. It doesn't strike me as a rage killing."

Court was recessed early that afternoon; Sheridan was scheduled to testify at another trial in San Bernardino. Two days later, his cross-examination resumed. At 10:12 A.M., he sat in the witness box. All the players were assembled. The mannequins of Rosie and Mandi were in place. But before the jury came in, defense attorney Hardy rose. "Your Honor," he said, "my client is not feeling good." Judge Yent suggested that since it was difficult to schedule Sheridan's appearances, they try to proceed until lunch. The bailiff called the jury in. Hardy continued his line of questioning about the number and nature of the wounds. He began to reexamine and scrutinize each of Rosie's wounds. The sensational and grim character of the killings had, after hours of testimony, affected everyone differently. Many jury members now struggled to pay attention. Jessielyn was back in court, making squiggles and spirals on a piece of paper. Debie glared at Underwood. A couple of courthouse regulars—two elderly women who were fascinated by murder trials—wandered in and out, speculating about how Underwood might have killed the two girls. As for the defendant himself, he was not faring well. He fidgeted more than he normally did. At noon, Hardy turned to Underwood. Underwood shook his head, looking haggard and depleted. Hardy turned back to Judge Yent. "Your Honor," Hardy said, "my client needs to go to the infirmary." Judge Yent said that court would recess for the day.

When Hardy resumed his cross-examination of Sheridan a few days later, the testimony ascribed yet more fury to the killings. Regarding the wounds to Rosalie's upper torso, "There were so many," Sheridan said, "I couldn't possibly separate them internally. Three wounds were eight inches. They went through to the back. There were multiple fractures in the rib cage. There were several stab wounds to the heart and both lungs. Most wounds in this group were capable of causing death on their own . . . If Rosie were standing when the attack started, she would collapse quickly . . . Each attack lasted about three to five minutes per victim, until the heart stopped beating." When he presented his case, this was information that Hardy would use to dispute what the prosecution maintained was the time line of the murders.

On redirect, Bailey asked if there were reasons for the different angulations of the wounds. In other words, because some of the wounds differed, could it mean that more than one person, for example, waged the assault? Sheridan said that there were three possible reasons for the differences in wound type: the assailant was moving; the victim was moving; or the assailant's grasp on the knife changed. Did this mean that he could have made a mistake in counting the number of wounds? Bailey asked. Sheridan reiterated that by his count, each victim had been stabbed thirty-three times. "I thought it was unusual that the number of wounds was the same," he said.

Debie pulled up outside Rosie's apartment in the Camaro. Even though it was the day before Mandi's birthday, she wanted to surprise her now, hoping that the gift would lift her daughter's spirits and maybe even give her an idea of what to do on the big day. The door was closed and Debie knocked. Underwood heard it and ran to the back bedroom, closing the door behind him, and crouching next to Mandi. Debie knocked again. There was no answer. Debie cracked the door and took a peek in. Underwood picked up the knife and waited. Debie spotted

Mandi's purse on the dining-room table. "Mandi," she called out. "Maaaan." No answer. She figured Mandi was sleeping; the bedroom door was closed and she knew Mandi never went anywhere without her purse. "Rosie," she called. "Knock, knock . . . Anyone home?" No answer. Rosie must be sleeping, too; from the looks of it—the score sheets from a game of cards, the spent cans of Olde English, a couple of stray cigarette butts—the girls had been up pretty late. Otherwise, everything was fine; Debie even took comfort in the empty ashtrays, knowing that no matter how much partying had gone on, her daughter had done a little tidying up at evening's end. This was good: Mandi was on a hair trigger lately and Debie figured that she was finally getting some rest. Better to let her sleep in. Debie was glad that she saw no reason to barge in; an unannounced entrance, even to check on her own daughter, wasn't Debie's way. It wasn't the way of anyone who lived out here; that's why they were here. And besides, it was somebody else's house. Rosie was Mandi's friend, not hers. She couldn't just walk in without some kind of prior understanding. Debie closed the door and drove to work in Mandi's birthday Camaro. The drinking and the vending of drinks started early in the Mojave, and it was time to open up the Iron Gate.

Underwood waited until the sound of the V8 had evaporated, wiped the knife, and dropped it, stepping on it with his Armani loafer as he walked out of the bedroom and, once more, over Rosie's body in the hallway. He cracked the door, and saw no one. Outside, he dropped a bloody washcloth in a garbage can next to Rosie's washer and dryer. He checked his watch and headed into the morning.

Outside Las Vegas, Juanita and Tom Brown were on Interstate 40 with Shanelle, hauling a load of groceries to Twentynine

Palms and points along the way—Baker, Needles, Barstow, and Victorville. Juanita liked Baker because you could stop at the Greek diner, underneath the world's biggest thermometer, and get a buffalo burger and fries. She didn't like Barstow because of all the hooker motels you had to drive past on the way to Stater Brothers to make a delivery; it reminded her of Manila, and all the women who were stuck there. And the only good thing about Victorville was that you were two hours from Twentynine Palms. "Mama," Shanelle said as they headed south on Route 247, the serpentine stretch that had taken Debie and her family to their new home, "I miss Mommy." She was sitting on Tom's lap. Juanita told Shanelle that she would be home soon, and she steered the truck into the passing lane, accelerating to overtake a caravan of Marine vehicles. "Why don't you tell Mommy that you're on your way?" Tom reached to the floor of the cab and handed the little girl her toy phone. Shanelle took it and dialed.

Over the years of his incarceration, Valentine Underwood's lawyers had repeatedly said that their client was concerned about being in jail in the desert. "Desert" was code language for "rednecks." After spending a few days in the tank at Joshua Tree when first arrested, Underwood was transferred to the Rancho Cucamonga facility near San Bernardino, where he had been held for six years. Not technically in the desert, Rancho Cucamonga was—like all of Southern California—informed by it. In every sense, it was a long way from the mid-Atlantic coast, where Underwood grew up. Black people lived in the desert, but the backbeat was the loud and pumping music of the predominant race, the hard rock of the 1970s and 1980s, Foreigner doing "Dirty White Boy," not Tupac Shakur doing gangsta rap; the desert was not the street or even the suburbs; there may have been a lot of hiding places, but if you were black, there was no place to hide, you stood out in the desert, unlike white people, who kind of faded into the white sands of the Mojave and its glaring white sun. Underwood talked of racist treatment at the hands of his mostly white guards. He told his family of various run-ins with prison authorities which he felt were racially motivated on their part.

If Underwood felt a pervasive racism in jail, he must have felt it even more keenly as he traveled on the prison bus into the high desert, over the Cajon Pass, which seemed either to flood, ice over, or catch fire, depending on the time of year, and into Victorville. As the bus headed past the Best Western where Debie was staying and turned off Interstate 15 and made its way north on Civic Drive, he would pass signs which directed motorists to the Roy Rogers Museum, a few blocks away from the courthouse and Victorville's biggest tourist attraction. He would not have been able to spot the signs from the prison bus, but no matter: anyone who had looked at a chamber-of-commerce map of the area would know of the presence of the Roy Rogers Museum, anyone with nothing but time to read and listen and ponder the surroundings would hear of the place. If Underwood felt like a stranger in the desert before his arrest, surely he must now have felt in his gut that he did not belong here. For if ever an institution gave off the foul stench of white colonialism in America, the Roy Rogers Museum in Victorville, California, was it.

The museum is situated on a vast plot of dirt that lies on the northwest side of town, between the freeway and the plaza where the town and the county and the state administer justice. Walking to it from the courthouse, from across the flats in the shimmering dust of the heat, a visitor spots a large fake horse on its hind legs, frozen in a posture of defiance. As the approach concludes, the strange desert equine comes into full view; it's a cheesy plastic sculpture of Trigger, the famous palomino stallion that belonged to Roy Rogers, complete with extended genitals that are anatomically correct, and enormous. Trigger stands before a kind of prefab fort wherein the artifacts of his celebrated owner's life are enshrined, including Trigger himself, stuffed and saddled up.

The museum is home to other nonindigenous members of the animal kingdom, such as Bullet, the Wonder Dog. It displays mounted animal limbs, and items fashioned of animal parts, all taken by Rogers on one of his many hunting expeditions around the world. Next to Trigger—"the smartest horse in the movies"—is a stuffed pronghorn antelope. There are stuffed wolves and foxes from Alaska. A monkey has been made into a rug. A zebra leg supports a stool. There are hippo tusks and rhinoceros and elephant feet. There are heads of antelope, wildebeest, kudu, impala, gazelle, and baboon. Presiding over the territory are six members of the Masai tribe—badly painted portraits of happy natives, perhaps the only happy incarcerated black people in the high Mojave desert, home turf of Roy Rogers, America's favorite cowboy.

During his incarceration, Underwood had been engaged for a few months. Christine Wilson was big and wide, not unlike women who often befriend men behind bars. She had painted fingernails that had grown into long curlicues, almost like claws, and she had to make sure to use her fingertips when performing the usual everyday tasks, such as picking up a newspaper or opening her purse. But her nails clicked whenever her fingers moved across a surface. All in all, she took up a lot of space. One day, while visiting an inmate at the West Valley Correctional Center, Christine met Underwood. The pair struck up a conversation, and kept talking. He would call every night, collect, from the pay phone in his unit, and they would recount Bible stories, especially the one his mother had taught him about the persecution of the innocent from the Book of Daniel. The conversation would always last for twenty minutes, which was the limit for calls placed from Underwood's unit. Christine was always home to receive the calls, even though they were costing her a small fortune. She was convinced that she and Val were soul mates and a few times even brought her kids to visit him. When he asked her to marry him, she happily told the world about it, proclaiming that her man was innocent of the crimes with which he was charged. "He couldn't hurt no one," she said. "He's my big ol' teddy bear." She said he was behind bars because he was black. "It's the Bible that's helping him get through this."

While Underwood was undergoing questioning by the police, he had asked for a lawyer fifty-four times. Each request was denied or ignored. It was the worst case of a Miranda violation in the history of San Bernardino County, so bad that the most incriminating part of the police interview was barred as evidence in the murder trial. In it, after denying having ever met Mandi and Rosie, ever having been at Rosie's apartment, Underwood finally admitted to being at the scene of the crime on the night of the murders. His lawyers had fought for years to keep the admission from being heard by jurors. But at the trial, Underwood took the stand in his own defense and told the world what he had told Neely and Palacios after the layers had been peeled off during six hours of questioning. Yes, it was his blood in Rosie's apartment. Yes, it was his sperm inside Rosie and Mandi. But you see, Your Honor, it all had to do with race.

On December 11, 1997, Valentine Underwood wore a dark suit and white shirt provided by his team's investigator, instead of his high-security prison orange. As he was sworn in as a witness, he looked across the courtroom and into the gallery, smiling when he made eye contact with his mother Zimena and his brother Tracy. It was the first time since the trial had started that members of his family had attended. In fact, it was the first time since the trial had started that there had been any spectators at all behind the defense camp; Underwood and his fiancée had broken up long before, and attempts by an observer to reach her had resulted in a classic desert dead end—a temporarily, then permanently disconnected telephone. Zimena and Tracy had flown in from Maryland and were staying at the Holiday Inn in Victorville for the duration of Valentine's testimony. In contrast to Valentine, Tracy was well-spoken, polished, clearly the product of a different experience. Zimena was old, bent, and frail. Although Underwood had often telephoned her collect from the West Valley Correctional Center, they had seen each other just once or twice since his arrest in 1991. In Victorville, they had reunited in private, before court was in session. Now, as he looked at his mother from the witness stand, she forced a return gesture from her seat behind the defense table, a sad nod. Perhaps she tried to recall another time and wondered how her son had fallen into this vast desert sand trap. "He would bring kids over and give them a bath," she once told an observer as if delivering a eulogy. "He put clothes that he had outgrown in the car trunk in case he saw someone who needed them. He took in dogs. Once he had nineteen dogs. He used to go fishing and he would distribute the fish to the neighborhood. One time he saw a cat chasing a rat. He took the rat from the cat and got bit. The doctors said, 'Oh my gracious.' You talk about a heart. He liked everybody and everything. He never seen a barrier when it came to color. Whites got close with him so quick—more so than his own people. My house would be full of white people. He didn't like you to judge a person by his color. The first time I heard him speak about prejudice was when he was over in Saudi. They had him on the front. It seemed like blacks were placed in the most dangerous positions. He couldn't understand it. I thought he would get killed over there. One day I was in church praying. I had a feeling I had to go home and watch TV. This was when the ground war broke out. I turned it on and saw him! I couldn't believe it! The camera kept getting closer and closer and he was standing like him, but I still couldn't believe it. Then he called and said he talked to a CNN anchorman and it was him. That's why I have to serve God."

John Hardy prompted Valentine to tell his story. It was long, filled with embellishment, puffed with the kind of useless detail that is the hallmark of smoke screens, the hat-hanging device of the trapped. Breaks came and went. Both Debie and Zimena quickly decamped to different ends of the courthouse during these breaks, talking nervously with investigators and representatives. Sergeant Neely was worried: Underwood was a good liar, he said, he had spent his life fooling people. He knew it could happen again. So did Lydia Flores, who recounted the story of how Underwood had grabbed the phone from Tammy Watson and told the 911 operator that his girlfriend had become hysterical when she called to report the rape. Observers speculated about the impact of Underwood's testimony. The jurors were scowling, some said. No, said others, they were hanging on his every word. Two were sleeping. No, they were all wide-awake.

Yes, he had been out and about on the night before the murders, Underwood said. He had stopped at the Club Max and had a Courvoisier—"it's cognac," he explained, taking up time—and then went to visit the Samoans. On the way, he ran into Rosie and she invited him over. Rosie was cool, Underwood said; he had met her on the base and known her for a couple of months. Asked by his lawyer what happened between him and Rosie that night, he explained that he "performed oral sex." "It was nothing cheap," he added, "it was respectful." A little while later, Mandi arrived. They started talking and Rosie left. "I ain't got no ties with anyone," he said. Then, Underwood said, Mandi pulled his pants down and they had sex. As Underwood spoke, Debie forced herself to breathe deeply and rhythmically; she had promised herself she was not going to "go high and to the right." Afterward, he left. Hours later, after partying some more at the Club Max and in Wonder Valley, he returned to Rosie's at sunup to see if there was any more partying to be had. The door was cracked, he said. The radio was on, it was staticky. He called out for Rosie. There was no answer. He walked in and discovered the bodies, and checked to see if they were alive. He knelt over Rosie to hear a breath. He looked at her throat and said fuck. He moved on to Mandi, again checking for a breath. As he leaned in close, an air bubble or something gurgled out of her throat. It made a weird sound. Shocked, he hit his hand against the wall. The force of it opened up a cut he already had and he bled onto the paint and the *Handbook of Practical English* on the floor below. Regaining his composure, he wiped himself with a towel and left. "I seen other dead women in Saudi," he said, "during Spearhead. We seen a lot of bodies, people lying by the side of the road, dead from airstrikes, decomposed . . ." His voice trailed off. "I never planned on being confronted about finding the girls and leaving," he said. "If I say, 'Yeah, I found them and left,' what's the consequences?"

During his closing argument, Hardy suggested that being a black man in America makes for behavior that white people might not find reasonable. For instance, he elaborated, when a black man, especially a black man from the South such as his client, comes upon the scene of a crime, especially a murder that involves white women, flight is the only thing that makes sense. "Mr. Underwood fled because he is black," Hardy said. "He knew that if he reported the murders, he would be blamed for them."

Olonda Anderson, a pretty black girl who resembled Queen Latifah, was on her way to the grocery store on the morning of August 2, 1991. Shortly after ten, she spotted Underwood walking north on Adobe Road, in the direction of the Marine base. They had met a while before, in the parking lot of Benton Brothers, a local family-run furniture store that did a lot of business with Marines and other residents on fixed incomes. She pulled over and asked him if he wanted a ride. He said sure, and climbed into the VW Rabbit, putting her youngest son on his knee and high-fiving her other two kids. She asked him where he was last Saturday night. "Oh, jeez," he said, working the moment. "I knew I shouldn't have gotten in this car." She pressed for an answer: "We movin' now and I'm behind the wheel." He told her he was at the Club Max. "You know I made dinner," she replied. "It was all ready and waitin'." "I tried to call you," he said, "but the pay phone was out of order." "Bullshit," Olonda said. "No, for real," Underwood said. "Why don't you just move into the Club Max?" she asked. "I would if I could," he said. "Phyllis treats me right." "You the weakest man I ever met," she said. "You didn't think that for the past three months," he said. "Yeah, well," she said. She dropped Underwood at the 3/11 barracks. He thanked her for the ride, and said, "Olonda, can I ask you a question? I need another ride. I gotta get to Norton for the game. Coach left at nine and I missed him. I'll be right back, just gotta go get my uniform." "I can't do that," Olonda said. "I need some time to think about things." Underwood leaned over and kissed her on the cheek. "Well," he said, "I know your kids come first. See you all later." The kids waved good-bye. He walked to the barracks, past the World War II cannons with the plaques that memorialized them as "Hellfire" and "Damnation," and headed for his room.

A year after the murders, former Marine Billy Dunlop received a phone call from Valentine Underwood at his home in Gallipolis, Ohio. Subpoenaed by the prosecution to testify in Victorville,

Dunlop said that the caller identified himself as Valentine. "He said, 'Do you remember me?' " Dunlop recounted. "I said no. He reminded me that I had given him a ride to Norton Air Force Base last year. He informed me he had been charged with rape. He asked me if I remembered that we had left about ten-thirty A.M. I told him that was wrong, we had left around thirteen-thirty. He said, 'Are you sure?' I said yes. He wanted me to come to California and testify on his behalf. I never heard from him again."

A little while later, Underwood hitched a ride with a Marine in an old gray pickup. "Where you headed?" asked the young white boot with the Oklahoma accent. "Norton," Underwood said. "Got a game." The boot was impressed that Underwood was on the team. They drove south from the base and turned west on Highway 62, just yards away from the mutilated corpses at the apartment on Palo Verde. "Feels good to get out of town," Underwood said. "Yeah," said the boot. "I'm goin' down the hill for some R and R." It was a familiar drive for both of them, west on 62, the only way to get to places like Palm Springs or San Bernardino, cities that were an hour or so from the base but really far, far away. The young Marine slowed as they approached a ramshackle cottage on the south side of 62 with a sign that said RAVEN'S USED BOOKS. The owner of the store, Raven, a gnarly ex-biker, sat on the stoop in his leathers and sipped espresso from a tiny white cup. He prided himself on grinding his own coffee, and served it up to anyone who browsed for a while. A longtime local who had dropped out of the L.A. tattoo scene years before, Raven had developed quite a reputation in the Mojave as a collector of old books, and chances were, if you were looking for an esoteric, long-out-of-print work, he had it in the stacks somewhere. Often, Marines frequented the establishment, looking for a wide range of stuff. The boot honked and Raven waved back. "Cool guy," the boot said. "Ever meet him?" "Nah," Underwood said. "He has good sci-fi," the boot told him, now accelerating, and heading through a stretch of nothing but the occasional Joshua tree to the south, and to the north, chirpy Lions Club signs warning drivers to slow down.

About forty miles to the northwest, in Cabazon, home of the region's primary Indian bingo emporium, Underwood spotted a sign for the Nike outlet. It was outside one of those vast outlet malls that have erupted all over the desert in recent years, giant shopping dungeons that cater to consumers on fixed or otherwise low incomes while at the same time functioning by default as centers of nourishment and fun for their kids. "Hey, man," he said, "stop here. I need some shoes. I been seeing all the young guys with the shoes that you pump up, and they keep saying when are you gonna get some money to get you some Nikes, and I got my leather Cons, but I gotta get me some of these new shoes since we here now." Inside the outlet, Underwood tried them on. The clerk said that the price was fifty dollars. Underwood asked the clerk to hook him up. "These shoes usually cost about a hundred," he said. "Guess I lucked out."

At Norton Air Force Base, Underwood quickly headed for the gym, just in time to suit up for the game. "Yo, 'Wood," his teammates said, high-fiving the power forward as he hit the floor. "Where you been?" "Out and about," Underwood said. "Hey, Val," a black girl in her twenties called from the bleachers. "Let's play some hoop!" "All right, green," another black girl next to her said. They were both in Marine gear, part of a raucous leatherneck rooting section that had come from the Marine bases at Camp Pendleton, Twentynine Palms, and Barstow to watch their team take on the Air Force. The rivalry between the green and the blue was as fierce as they come. The

Corps was the scrappy, street-fighting team, reflecting its population of hard-core, what-you-see-is-what-you-get types. The Air Force had polish, finesse, reflecting its slightly more monied gene pool. Halfway through the first quarter, the Marines were up eleven to eight. Underwood had scored three times, twice in the paint and one three-pointer from uptown. He was doing pretty well, but knew he could have been doing better, should have been doing better, now that he had himself some shoes: every time he went in for a hoop, his left hand started quivering. That was why he had just missed an easy layup, causing his coach to jump up and shout, "What the fuck, 'Wood?" Underwood shrugged sorry and shook off the mistake, running back up the court to defend against blue. Blue sank a score and he grabbed the rebound, striding back way ahead of the Air Force guards to try for another. But this time, when he leaped up for a slam dunk, the fingers on his left hand started shaking. He felt the ball quiver ever so slightly, then watched it fall away as the cheers of the Marine fans instantly faded to silence. The coach called a time-out and Underwood walked to the bench, head down, trying to shake off what had just happened. "Coach," he said, "something's wrong with my hand." He put forth his left hand and it was shaking. "Go see Doc," the coach said. Underwood grabbed a towel, wrapped it around his neck, and proceeded down the bench for the medical corpsman. "Doc," he said. "Check this out." The corpsman examined his nervous left hand, top first, then palm up. He noted a cut and was surprised at its depth. About three inches long and deep enough to expose tendons, it ran from between Underwood's thumb and index finger almost to the wrist. "I cut it on a razor reachin' my hand into a barrel," Underwood said, explaining that early that morning, while on latrine duty at the barracks, he had reached into a garbage can and pulled out a Bic razor, catching the blade deep in his palm. The corpsman sent him to the clinic. When

Underwood got there, his hand was shaking so much that he asked another Marine to fill out his medical forms. "This is pretty nasty," the doctor told him after a twenty-minute wait. "Looks like a lacerated tendon." "It was healing up pretty good," Underwood said, "but then I guess it opened up again during the game." The doctor asked him who was going to win. "Now that I'm out," Underwood said, smiling, "looks like it's gonna be Norton." The doctor bandaged up the wound and applied a cast. "You're good to go," he said.

From a pay phone outside the clinic, Underwood made a call. "Hey, Cheryl," he said, "guess who's in town?" After arranging a date with an old flame who lived on base, he returned to the locker room, showered, and changed. Later, he and Cheryl had a few drinks. The following evening, he was back at the NCO club with his teammates. Not wanting to draw attention to his hand, he had removed the cast. But his injury was the talk of the table. A teammate suggested that if he planned to cut his hand on a Bic razor again, please make sure to do it during the off-season. "Sorry, man," Underwood said, knowing that the team would have to withstand months' worth of derision over the embarrassing loss to Norton. "It was lame, I'll be the first to admit it." "Hey, 'Wood," the coach said. "Got some advice. You should put the cast back on because the ladies like it."

"Have you ever taken a polygraph before?" Neely asked Underwood during the initial police interview. Once, Underwood replied. Neely asked him what it was for. "That was, that's when my ex-girlfriend said that I beat her up and shit," Underwood said. "It was in Maryland, years ago, in '86 I think." Neely asked Underwood what other criminal things he had been involved in, in addition to the assault and the rape that he had mentioned. "Okay, sir," Underwood said, "the assault that I was involved in was my ex-girlfriend. I had, ah, if you look at the report, this was all fucked up. Me and her was working together, we was fucking around with each other, they also tried to put a charge on me in Binghamton, New York, that I could have sued them on. I had, my people told me I could have sued them, I said no, fuck it because they held me up there, they called me nigger, they treated me like a piece of shit, and I gave them money, ah, I think a few thousand dollars that it took me a long time to get back. They gave it back to me. Ah, a bunch of bullshit, and then, that's why I'm saying, sir, that's why when, that when this girl tried to say that rape shit, that shit fucked with me bad, and I told you and I answered you I said why I end up having this shit, try to happen to me once before. And I kept asking myself why in the fuck is it happening to me."

Neely followed up on the Binghamton incident. Underwood said, "I worked for a company called Atech. I was doing, ah, engineer consulting. I was going around, ah, doing polyurethane tests, infrared tests and everything like that. 'Cause my minor was civil engineering, ah, ah, I met this girl she was in there partying, we was, I was in town, and, ah, the only thing that had saved me, to date, the people that work at the hotel knew her and knew her background that she was scamming people. Right?

"She had put some shit in my drink and had me kind of disoriented. I'm wondering what the fuck's going on. Me and her fooled around, I had grabbed hold to her and told her that I was real fucking dizzy. And, ah, I told her, I said you know I'm going to get your ass. Next thing you know the police comes in and they taking me in, and saying that she said that I had, I had sex with her and that she didn't comply with it . . . They had me over the Fourth of July holiday. They wouldn't let me, they didn't, they didn't read me my rights, they wouldn't let me see nobody and nothing, they kept me in jail, wouldn't let me make a phone call for a few days, they broke all my fucking rights."

"Any other time you been arrested?" Neely asked.

"When my, when my, ah, the ex—I was, this girl I was dating," Underwood replied, "said that I assaulted her."

"What kind of charge was that?" Neely asked.

"I'm the one that called the fucking police," Underwood said. "She said that I had beat her up, because, okay, she then said me and her had screwed around. She was telling me about back home, sir, she was also messing with this guy that was a drug dealer. I didn't mind, sir, 'cause she was giving me money. As she was working with me at a vocational center at a religious foundation. And, ah, she said, that me and her, ah, me and her messed around that day and, ah, she told me to leave 'cause he was coming. There's a whole lot of shit. I can't remember everything."

"So," Neely asked, "was the charge an assault or—?"

"It was an assault," Underwood said. "She said I assaulted her and she tr—and, ah, she said that I had ra—ah, raped her. She said a bunch of shit, sir."

"How come," Neely asked, "you keep getting in all this, this hot water with women?"

"Women?" Underwood replied. "I don't know. I don't know . . . The bad part is I got a girl at home and she is a good girl, and I said I'm not going to fuck that up. Me and her has been talking about getting married, I said I'm not going to fuck that up."

Debie had been calling Mandi all day, with no answer. Shortly before 8:00 P.M. she pulled up in the birthday Camaro. Cops were putting up the last bit of yellow tape denoting a crime scene. "Oh, my God," Debie said. "Oh, my God." She did not know what happened. She rushed through the tape, toward the apartment, but was stopped by a huge paramedic. "Debie," he said, "Mandi's dead." Debie started to faint into his arms. The paramedic applied an apparatus to her left arm. A cop held Debie on the right. "She's going into shock," someone said. Debie ripped the apparatus off her arm and, still restrained by cops, flailed and heaved herself toward the apartment entrance until she ran out of steam.

A couple of minutes later, Detective Norm Parent arrived, along with a pair of criminalists. They donned protective footwear and entered the apartment. Parent—a twenty-year veteran of homicide investigations—noticed the odor of death in the air. They set up high-power halogen lights to illuminate the residence. The light rendered the look of life on the edge even more desperate than it actually was, exposing without any relief or shading the facts of this dwelling—a little basket of layaway bills from Wal-Mart and pay stubs from Texaco, a collection of gangsta rap CDs stacked up next to the boom box, a couple of yellowing envelopes from Batangas on the kitchen counter, a plastic bowl with a few remaining Cheez-Its. The investigators began to note and gather evidence, proceeding to the back bedroom, hunting for the start of the trail. And there it began. There was a bloodstained tissue on the floor. There were items of clothing strewn about. There was a black sock with a stain that was logged in as "white" and "crusty." There was a bottle of a sexual lubricant called Motion Lotion (mistakenly logged in as "Emotion Lotion"). There were several cardboard boxes scattered around the room. Next to one box was a large knife—and Mandi, with a stiffened leg atop the container. Parent spotted bloodstains on the carpet throughout the bedroom, and then, on the west wall, a bloody handprint, a fresh desert pictograph indicating . . . what? Was this the bold signature of the killer, could someone really have left such a disclosure behind? There were more blood splatterings on the wall, on one of Rosie's fashion magazines, and on the cover of *A Handbook of Practical English,* which she had been studying in her spare time.

The investigators came out of the bedroom, into the hallway, and were stopped by Rosie's body. To examine it and gather evidence around it, they had to back into a small closet across from the bathroom and work out of it so they wouldn't disturb the corpse. Peering over the body, Parent noticed blood splatters all over the blue walls, the floor, the sink, on a pink blanket in the bathroom, on the closet wall, and on the underlying hallway carpet. As they worked their way out of the closet and around Rosie's body and into the living room/kitchen area, Parent jotted down the last signs of life, saw from a different perspective the signs that had indicated to Debie when she stopped by that morning that everything was fine: beer cans on the kitchen table, playing cards, a score sheet on a notepad, a purse.

And then there were the bodies of Mandi and Rosie. As the police deciphering began, the voices of these young girls who mothered, danced with, cooked for, gave sexual favors to, raised, guided, took care of their friends suddenly began to scream loud enough for the outside world to hear and record. Now everything about their lives was under scrutiny: hair was plucked from their heads, bellies, shoulders, thighs, arms, wounds; fingernails were extracted, their vaginas were swabbed, fiber was taken from a slash to Rosie's chest, blood was suctioned from their torsos.

Six hours after investigators arrived, the bloodiest crime scene in the history of Twentynine Palms had been cataloged and collected, the physical evidence of a sad and tawdry Mojave tale carted off without ceremony to a police locker. The apartment had been a treasure trove of clues, yielding much more than the typical crime scene. Detective Parent figured that if the killer or killers could be caught before the trail grew cold, the case would get to trial in one, maybe two, at the outside, three years.

At one in the morning of August 4, 1991, the coroner's van arrived to take the bodies of Mandi and Rosie away from the apartment that was a bloody wreck of life's milestones. It was Mandi's sixteenth birthday. In a month, Rosie was to turn twenty-one. In two months, the killer would turn thirty. Debie was still outside with the Camaro as the paramedics carried the girls into the van. Because she had stopped at the crime scene the morning before the bodies were discovered, she was now officially a witness who would be called at trial. As such, she was barred by police from accompanying her daughter to the morgue. A detective told her she could go later to claim the body. As it turned out, she would not get to say good-bye to her daughter for four days.

Once again shut out from one of life's expected graces, Debie stood alone in the night, wailing under the stars, as the cops stood guard outside the flimsy door marked "7," next to the double-decker washer and dryer that Rosie was still paying off. Sometime during the evening, Krisinda got the news at Pizza Hut, where she was filling in for a friend on the night shift. Jason, however, knew about it long before his mother or his sister, sensed that the unspeakable had happened. He had been hanging out with friends at a house in Rosie's neighborhood. All night long and into the early morning, they had heard the sirens coming and going, watched the flashing red lights. "Let's go see what's up," one of the boys kept saying. Jason said no. "Come on," the others urged. They always tailgated the cops around town; there wasn't much else to do, and it usually had something to do with someone they knew. But Jason just shook his head. He had an upset stomach, he said, and his friends figured it was pretty bad; he was doubled over in pain, curled up on the floor for hours and rocking like a little kid. Whatever bad thing had occurred, he knew that it involved Mandi.

PART FOUR

Tribulations

Now the moon is almost hidden
The stars are beginning to hide
The fortunetelling lady
Has even taken all her things inside
All except for Cain and Abel
And the hunchback of Notre Dame

Bob Dylan, "Desolation Row"

Lay her i' the earth:
And from her fair and unpolluted flesh
May violets spring!

William Shakespeare, *Hamlet*

She is a dynamite dreamboat
A load of atom bombs
The lady from Twentynine Palms

Allie Wrubel, popular World War II song,
"The Lady from Twentynine Palms"

On the day of the murders, Tina Herrera had been calling Rosie's apartment every hour on the hour, sometimes in between. Although the two had always celebrated their birthdays together, Mandi had told Tina during the weeks prior to August 4 that she just was not in a party mood this year. The implication was that each girl would have to turn sixteen in her own way, which Tina did not understand because she and Mandi had so often talked about their plans for the future. Their main plan had to do with moving to Santa Cruz after high school, to live with Tina's grandmother and, since both loved swimming, swim in the ocean every day, as long as it took to wash off the desert. From her bedroom inside the family home on Bedouin Avenue just outside of Joshua Tree National Monument, Tina kept dialing and dialing the number Mandi had given her. Periodically, her mother would peek in and ask her daughter what she wanted to do for her birthday; after all, it was tomorrow, and she still had not made any plans, insisting that she first hook up with Mandi. Even if

Mandi didn't want to have a party, she at least wanted to bring her the jumbo bag of M&M's she had bought on special at the Kmart in Yucca Valley, to let Mandi know they'd always be friends, no matter what, it would be the LBG forever, just like Mandi's poems said. But the sunlight vanished and Tina finally stopped dialing Rosie's, hanging up for the evening with a queasy feeling. She came out of her bedroom. Her mother was in the kitchen, baking a coconut cake for her birthday, puffing on a cigarette as she blended the ingredients, banging around a little more than usual. Debie had called, looking for Mandi, figuring, hoping, that she might have checked in with Tina regarding her birthday. Tina's mother, a slim and nervous woman who was calmed some by the wide-open space of the desert, whose free-floating anxiety had only the scenery to latch on to out here, said she had a bad feeling about Mandi, echoing her daughter's fears. Tina asked if she could go to Rosie's to see what was up, even though her mother never let her go there because of all the Marines who were regular visitors. Her mother said no, she was not making an exception. So Tina and her mother stayed up all night, and when it was time to decorate the cake, Mrs. Herrera squeezed out SWEET SIXTEEN, TINA AND MANDI in green sugar paste on top of the coconut cake and Tina made a heart in the shape of M&M's around the message and then they finally went to bed as dawn broke over the Mojave on this important day.

Rose Powell had not seen Debie since the day she had left Oakland for the desert. Rose barely knew Mandi, but now she was coming to Twentynine Palms to tell her granddaughter good-bye and to try to comfort her daughter. She had hooked up with a new beau who lived in Reno, and he was driving her down through the desert and into California. Outside San Bernardino, the oil derricks puffed toxins across the landscape. A black Nissan cruised by, jacked up on monster wheels and carting a small crew of dark gangbangers in blue bandannas. From the car came a quick blast of rap music and air-conditioning; on this white-hot Mojave morning, Rose looked at the car and then looked away as the bangers flashed the Crip sign, a downturned wrist with the thumb and forefinger curled in a C shape.

Outside the mortuary in a strip mall at the corner of Desert Queen and Highway 62, there was a crowd of two hundred or so, of all ages. They could have been there for a special at the pizza place or maybe Stihl was staging some sort of promotion for chain saws at the hardware store—the company was always recruiting contestants for extreme logging events that they staged in rural outposts around the country. When Rose got out of the car, she had no idea that the crowd was standing in line for her granddaughter's funeral. She noticed the black Nissan amid an impressive array of vehicles—lots of other shiny, new cars with dark windows, a fleet of Harleys, some old Chevys and Pontiacs which recalled arguments she had heard men having for as long as she could remember: "I'm a Ford man"; "Not me, I'm GM all the way." She looked for Debie, and started to make her way to the nondescript storefront that was the entrance to the funeral home. Overhearing expressions of grief, she realized everyone was here for Mandi, and she moved with Mandi's friends through a phalanx of Hell's Angels who had come to pay respects to a brother's goddaughter and perform services in crowd control, a field in which they excelled, especially when life-and-death rituals were involved.

Into the home walked many of those who dwelled in Mandi's magic circle—her Samoan friends, her friends in the Crips, her friends in the Bloods, single mothers and their kids, schoolmates, girls on her swim team, Jason and his buddies, the Lunch Box Gang in their freshly laundered LBG T-shirts, Rosie's mother Juanita Brown and her sister Jessielyn Ortega. They stopped and signed their names in the guest book, many with the carefully crafted signatures of those who find identity in the few places available to them, making sure to add their nicknames or affiliation. Juanita signed in, for herself and Rosie's daughter Shanelle, although Shanelle was not there. There was Vanessa Roberts, Dana Phelps, Sunshine Carrol, Jodie McKittrick, Mad Dog, Tony Baynes, Smokey Chaplain, James Morgan (Little man), Tam-me Barnett (LBG), James E. Morris III, Michael Givens, Jr., Debie's brother Darrel McMaster, Debie's father C. D. McMaster, Desiree R., Mysti Thompson, Ebony, Erin, Eric, and Danielle Turner. For a while, Debie greeted each mourner, taking in the full impact of her daughter's life, realizing in her gut what she had long joked about—that it was "like a mini-UN," different in this way from her own life, with so much more promise. But when Debie's family arrived, she froze; the words which came easily and provided a sanctuary for the grieving—"Thank you for coming," "Mandi loved you, too"—stopped flowing; in fact, she could not speak at all and fumbled in her purse for a cigarette. She was out. "Have one of mine," Rose said, stepping forward to try a connection with her daughter, and offering a Newport. Debie thanked her and lit up. Rose lit up, too. Neither said a word about the history of women in their family; they smoked and found solace on the exhale.

Inside, "Stairway to Heaven" filled the room. A blown-up photo of Mandi—her ninth-grade yearbook photo—rested atop an old Formica table. Corky sat before the altar. Debie and Rose took seats in the front row, next to Krisinda and Jason. Debie kept a keen eye on the mourners, wondering if her daughter's killer had sneaked in (and who he was), agonizing over whether her daughter's father would walk in, a day late and a million dollars short. He did, along with his second wife. Except for the one phone call she had made to Max on Mandi's fifteenth birthday, Debie had had little contact with him since he tried to keep the children in Texas. He gave her a nod as she made eye contact. "You son of a bitch," she started to say. Her mother intervened. "Baby, sssh," she said. "You got other things on your mind."

Pastor Frank Kernodle began the service. His remarks were brief and generic, for he did not know Mandi, and there were many who did and they all had something to say. Lebow Pritchett, a pudgy Crip with a quick smile and a temper to match, read a note from Kevin James, who was in jail, detained as a suspect in Mandi's murder. "I wish we didn't have to say good-bye like this," Kevin wrote. "Don't worry, Mandi, I'll get the bastard, wherever he is." The girls in the Lunch Box Gang called out their agreement and so did the guys from the black Nissan and there were many revival-tent echoes of "You got that right" and "We love you, Mandi." They all knew Kevin was not responsible for the murder of their friend, although he certainly was capable of violence. Seven-year-old Bo Barnett placed a bunch of wildflowers in front of Mandi's picture. "Mandi took me to play miniature golf before it went out of business," he said. Each member of the LBG recited a poem she had written in Mandi's honor. Tam-me called hers "This Girl." Mimi read "Tears." Tina, with whom Mandi had run away in fifth grade, remembered Mandi with "Friends." Lydia read "Best Friends." Even Janita, Kevin James's current girlfriend, now eight months pregnant, stood up to honor her friend, reading one of Mandi's own poems, called "Tomorrow." Lebow asked the mourners to rise for a final tribute. A song by her favorite group, Boyz II Men, was played. It was "It's So Hard to Say Goodbye to Yesterday."

As the guests left, they signed a memorial scroll. "Watch over me Girly Girl," someone wrote. Mandi's brother wrote "What's up?" and signed it "Jayster." "Amanda," said another friend, in the unintentional blank verse of the typical teenager, "so many times i want to tell you so many things, you were the only one who understood, i care for you so much i wish you could be here with us still i have to tell you so much stuff that has happened with us, but I'm sure you've seen watch us Mandi and Rosie, watch us good. I ♥ you both. Rest in peace." And then, a warning: ". . . the asshole will die Amanda he's gonna burn in hell. I promise . . ." The messages on the scroll were not about shared memories, nor did they bear wishes for peace on the other side. They were about protection and revenge. This is how it was, still is, on the modern frontier: in one of the most heavily armed towns in the world, kids were scared and took care of each other if they could, and tried to mete out justice if they couldn't.

Two weeks later, Debie went to the mortuary and picked up Mandi's ashes. She had not been able to do this earlier, and in fact Mandi's urn was not at the funeral; the act was too final. Now Debie carefully placed the urn in the blue birthday Camaro.

In the late-afternoon sun, mother and daughter drove deep into the Mojave, past the giant boulders at which they had marveled, past the Joshua trees that had hinted of all the wonder in the world, to their favorite spot. From the car Debie looked across the desert to the beginning and end of time. "Baby, I fucked up," she said. "Big time." Her tears came and went quickly, this dry zone soaking up, welcoming, needing the grief. Debie smoked a joint and they stayed for a while. Then she took Mandi home.

Rosalie Ortega went home, too. Jessielyn raised money among the Filipino community on the Marine base so that her sister's remains could be flown back to Batangas, accompanied by her immediate family. Dozens of relatives gathered at the family plot for a graveside service. What had happened in the Promised Land? they wondered. How had one of their own traveled so far only to be beaten down so low? Rosie was too good for this world, a priest told the wailing mourners. God wanted her for his garden. After the final blessing, Jessielyn lit up a cigarette, took a long drag, then placed it atop her sister's coffin—a last, satisfying Kool for Rosalie. Shortly after Rosie's funeral, Juanita Brown's older sister—Rosalie's aunt—died unexpectedly of a heart attack. Juanita's brother collapsed and died, too, of heart failure. Just in their forties, they were healthy, like Rosie; on that night in 1991, it would seem that it was not just Rosie who was killed, but that the Ortega family was sliced to the heart.

Among the establishment of Twentynine Palms, a view of the murders was circulating that was far different from the views making the rounds elsewhere. "This is why I steer clear of the trash in town," said a local merchant at the bar of the Inn at Twentynine Palms. The bar caters to well-to-do locals, Marines—generally those who are commissioned—who want to impress their girlfriends, and tourists who can afford the Inn's prices, a group that, generally speaking, is proud of their escape from humble origin and wants nothing to do with reminders of it. "All these people on welfare, they can't take care of themselves," he continued, expressing an ancient American gripe. "They expect handouts from Uncle Sam; why can't they just get a job and better themselves?" The tourists at the Inn had not heard about the murders. They rarely ventured into the town itself, where the murders happened, and the murders were not covered in the glossy periodicals from which they got their information.

But the news was on the front page of the other local newspaper, *The Hi-Desert Star,* available at kiosks in town, under the banner headline MURDERS IN TWENTYNINE PALMS. A group of Mandi's friends lingered outside Rosie's apartment, still yellow-taped as a crime scene, clutching the latest edition of the paper, rereading it for answers, keeping a vigil. Although some of them lived in households where violence was routine, the crime that took Mandi and Rosie was way too grisly to dismiss. In fact, they knew that if they had been at Rosie's that night, they would have been victims. Of whom or what they were not sure, but they were frightened that something so evil had struck so close. Lydia, with her arm in a cast, was the last of the LBG to have had a conversation with Mandi. She told Tina that Mandi asked her to meet her at Rosie's that night. But she and a friend of the family had been fooling around, and somehow her arm broke. Like Mandi on that night, Tina read between the lines of Lydia's story and said nothing, wondering if an act of violence at home had kept Lydia from suffering a greater violence elsewhere. But tough-girl Lydia did not regard the incident as a lifesaver. She blamed it for keeping her from a friend in time of need, and said that she would have kicked their ass, whoever the murderers were. "Motherfuckers!" she screamed for all the desert to hear, picking up a handful of rocks and throwing them at a long-abused mailbox.

The disappearing sun marked an end to a day of grieving, and as the girls walked south on Palo Verde toward Highway 62, they talked about who might have killed Mandi and Rosie. They had heard all of the rumors and didn't buy the ones that involved the usual suspects—their friends, "the trash in town." Yes, some of their friends had police records. Yes, some were violent. But not violent like this. And in a town where passions ran high, Mandi and Rosie were generally a calming presence, although they would not avoid a fight if required. The girls could come to only one conclusion. Who were the people who caused the most trouble in town? Who was always stirring up hate and discontent and getting away with it? Who was at the Motel 6 every payday, shacking up with underage girls and beating them up and the local authorities always looking the other way? No doubt about it—it was the Marines. Things were really out of control now, Lydia worried. Because this time, they probably really would get away with murder.

Debie checked in for her evening shift at the Iron Gate a few weeks after Mandi was killed. She brought Mandi's urn and Corky and put Mandi up on a shelf behind the bar, next to a few faded red, white, and blue streamers from the Gulf War victory party, surrounded by all the bar-joke bumper stickers, most of which suggested that the only thing women wanted was to get fucked. She did not yet know all the details of her daughter's death, that she was a war hero of sorts, a victim of collateral damage, a casualty of the never-ending Marine war against female civilians. "Watch over Mandi," Debie told Corky, "sit here." He sat behind the bar, just under the shelf with the urn.

A few Marines had settled in for an evening's bender. Each was already on his second individual pitcher of beer. One of the Marines asked Debie what was in the urn. Another Marine joked that it was probably a pet's ashes. Debie told them that it was her daughter. "Your kid's in there?" the first Marine said. Debie said yes, that was correct, and asked him if he had any problems because of it. "Well, this ain't no mortuary," he said, and cracked up. Then the other Marines started laughing, too. It was too grim to take seriously. "She's got her kid in there," the first Marine said, to the other bar patrons, "can you believe it?" A couple of patrons eyed the urn, then got up and left. Debie's boss took her aside and asked her to take Mandi home, suggesting that it was too loud in the bar and this was not a place where Mandi could rest in peace. "Are you saying I'm bad for business?" Debie asked. "Because you know I got a following. Half the people in here are here because of me." The Marines recovered from their fit of laughter, swilled some more beer directly from the pitchers, then got maudlin. "You know, that urn gives me the creeps," the first Marine said. "We ran over some I-raqis in Saudi, why I gotta come back to this shit?" He got up, threw some cash on the bar, and left, followed by his buddies. Debie's boss asked her again to take Mandi home. Debie replied that from now on, Mandi would stay with her. The bar was a good place for Mandi, Debie said. Mandi was a people person, and besides, she liked to dance. "No one can dance to 'Layla,' " Debie's boss said, which was playing yet again on the jukebox. Debie scooped some quarters from the bar counter, ran to the box, and threw a few in. "Well, we can fix that, can't we?" Eric Clapton ran out of steam and the needle plunked down on another disc, "Wild Thing" by Tone Loc. "Mandi loved to do the wild thing," Debie said. She hopped on a table and started to dance, a desert banshee gyrating to banish the thing that had once again caught up with her family. Earlier in the day, a speed freak had wandered across Highway 62, tweaked, flailing, finally lying down in the middle of the road until he was picked up by paramedics after causing a Marine on his way to Laughlin for the weekend to roll his new Mustang so as to avoid running over the desert tweaker. No one paid it any notice, especially the paramedics who picked him up; it was a routine start to another desert day. But even to the most jaded customers of the Iron Gate, the regular two-pitchers-by-sundown crowd, Debie's behavior was cause for concern. One of them suggested that the heat was making her loco. Another offered the kind of small-town analysis that proscribes rituals such as mourning: maybe she was tweaked, after all, that's who lived in Twentynine Palms, case in point, yours truly, ha-ha, and besides, if you've just lost a loved one, how can you come to work? Debie's boss asked her if she wanted to take the rest of the night off. Debie declined, explaining that the bar's sound system was better than hers, so why would she want to go home? "Someone turn that baby up," she said. "Mandi loved this rap shit."

. . .

Since the night Valentine Underwood had attacked her, Tammy Watson had rarely left her house. Although she had been told that the Corps would deal appropriately with Underwood, she was afraid—of anyone and everyone, especially the Marines in this military town. And there was the shame. A few days after she had told her father about the incident, he began to suggest—over the phone, from North Carolina—that perhaps she had been hanging around the barracks too often, why was she always chasing after Marines, and didn't she realize how running around town could make things really difficult for a guy like him—the subtext suggesting that her behavior could cause trouble for a high-ranking black NCO such as himself. Seventeen-year-old Tammy did not know what to say. Having grown up as a child of a Marine, she was very mindful of the Marine slogan, "Semper Fi." It wasn't that members of her family, or the extended family of Marines in general, would say "Semper Fi" to each other when they got up in the morning, or passed each other on the street—although like all Marine towns, Twentynine Palms sported its share of "Semper Fi" bumper stickers and T-shirts. Saying "Semper Fi" wasn't necessary, it was almost redundant, for it was understood among Corpsmen and women and their spouses and children that Marines were "always faithful"—first to their unit, then to the Corps, then to country, and finally to God. There was no room for conflict in this code, and how could it be otherwise? America's premier fighting entity—first into a war zone, last to exit, never to leave a body on the field—maintained loyalty to those in its ranks and those it served to guarantee its very survival. When Tammy told her father that she had been raped by a Marine, she knew she was asking him to put loyalty to her above loyalty to the Corps, to break the ancient code of the warrior's silence. But she wasn't prepared for the emotional battering she would suffer as a result.

Nor was she prepared for the front page of *The Desert Trail* several weeks after the attack. She was still living in Desert Hot Springs, about twenty minutes southwest of Twentynine Palms, at a lower elevation of the desert, in fact at the intersection of two deserts—the Colorado and the Mojave. Nobody knew exactly where the two officially commingled; old-timers believed that the confluence of the two, combined with a third element, the underground springs which fed the town's many "spa-tels," gave the town mysterious properties, often described as "yeah, we got some kind of vortex going on here." Along with the usual round of desert denizens, all manner of healers have flourished in Desert Hot Springs, servicing all manner of pilgrim. A singing psychic facialist serviced a small, devoted coterie of toilers in the entertainment industry who swore their lives were changed forever when Lois Blackhill applied her masques with the secret desert ingredients and simultaneously channeled the operatic voice of an ancient Chinese spirit named Omra. An iridologist in a clinic under a grove of palm trees looked in your eyes and told you what ailed you, for a first-time fee of $225, then $75 for subsequent visits. A latter-day disciple of Christ channeled Christ energy to anyone who wandered into her motel and asked for a room, help, a dip in the waters. Tammy Watson, religious by upbringing and belief, did not solicit the aid of any of the local healers; like many who dwelled in the desert permanently, she could not afford their prices, and neither had she heard about most of them. She, and others like her, were the kind of pilgrim who was truly lost, consumed by cycles of violence which would hardly be quelled by a facial or bathed away by an hour in the town's hot springs with its mysterious healing properties. Yet Tammy was happy to be living in Desert Hot Springs, especially after the rape in Joshua Tree. She didn't have to deal with the military every day, like people who lived in Twentynine Palms, especially since Underwood had been phoning her friends, asking if they knew where she was. Desert Hot Springs was closer to Palm Springs and its twenty-two golf courses and its renovated fifties-style bungalows and its avenues named after Bob Hope and Gerald Ford, and therefore more in its orbit than Twentynine Palms, and that meant she was that much closer to having a toehold on respectability. So one August morning, six weeks after the rape, she ventured out of her house, walked down Hacienda, and turned south on Palm, the town's main drag. She walked past the fire station, the post office, the library, several chiropractors' clinics, a health-food store, a Hungarian restaurant known across the desert for its superb red cabbage, a video store, a pizza place, and entered the old diner at the corner of Third and Palm. Bev the waitress was glad to see her and inquired as to where she had been lately. Tammy shrugged it off, took her usual counter seat, lit up a cigarette, and ordered a coffee. As she sipped, she spotted a copy of *The Desert Trail* lying on the counter. The Twentynine Palms newspaper was not generally distributed in Desert Hot Springs, as Desert Hot Springs had its own local paper; obviously a patron had left it. She reached over and picked it up. The headline, above the fold, said, MARINE HELD IN SLAYINGS. There was a picture of Valentine Underwood, the man who had attacked her, along with an article that told of Mandi Scott and Rosalie Ortega, two girls Tammy did not know but to whom she now felt all too close, and how they were stabbed multiple times, allegedly by the Gulf War veteran. The news struck her like a kick to the stomach; she started hyperventilating as she reread it and ran to the bathroom, where she was overcome by an attack of the dry heaves. The memory of her conversation with her father repeated itself: the Marines will deal with Underwood, drop the charges, let the Corps deal with its own. Now, added to the shame of her rape, came the blame that she assigned herself for the two murders. If only she had had the strength to take Underwood to court, she thought. If only she had followed up with her father to make sure Underwood had been disciplined by the Corps. If only she had contacted the *Los Angeles Times*. "Everything all right in there?" came a voice outside the bathroom door where Tammy crouched on her knees, cradling the toilet. "Your eggs're ready."

Two coke freaks from a tropical island landed in Twentynine Palms one day, tired of the South Pacific and looking to start over in the desert. Married for fifteen years, the couple known as Sam and Michelle had pooled what little cash they hadn't blown on drugs and rented one of the various empty buildings on Adobe Road, a few blocks north of Highway 62. They went to work right away—scraping the old landlord-white paint off the walls, repainting them with lush pastels, wiring an overhead track-lighting system, and, their last task, installing the twenty video games they had rented from the area's dealer of such things. And thus did their dream erupt in the Mojave; the Silver Quarter they called it, Vegas-style. It was the only arcade in town, and it quickly attracted the customers it had sought—kids who had nowhere else to go, nothing else to do—becoming a desperately needed de facto day care. But then followed the rest of the young parade: Crips, Bloods, Samoans, Latino outlaws—*vatos*—Marines, and the young girls who trailed them everywhere. It wasn't long before local cops and the NCIS started cruising by, just to remind everybody who had the upper hand in the local hierarchy of hardware.

Just as volatile situations have a way of erupting, so also are they magnets for peacekeepers. Enter a local who called himself B. T., a six-foot-and-then-some 170-pound ex-Marine, a tired warrior with leading-man looks who when pressed by those who got to know him over time revealed that he was the son of a man who was in jail for manslaughter ("where I don't know") and claimed direct descent from Geronimo and Billy the Kid—a violent heritage of which he was ashamed, a blood legacy which even if invented might as well have been true: every waking moment B. T. appeared to have to remind himself not to explode, and the flat expanse of the desert seemed to suck out the volatility. "I heard you need some help," he said to Sam and Michelle when he walked in to the Silver Quarter one summer morning after breakfast at the Jelly Donut and an all-night drunk (although only his bloodshot eyes gave any indication of his being anything but clean and sober). "I'm good at running things. Kids like me," he continued, gesturing to a few in the arcade who returned the greeting with a "what's up?" The thirty-two-year-old local offered his résumé. He had run away at seventeen from a small town in Michigan to join the Corps, he explained. He had worked his way up to staff sergeant and his MS—military specialty—was admin; once he was in charge of a million-dollar budget. After ten years, the Marines threw him out because he couldn't stop drinking. If he had been able to stay off the stuff for just a few more months, he said, he could have retired with full benefits. Although the court-martial stripped him of his rank and retirement earnings, it was obvious that he still considered himself a Marine and wanted to join some team, somewhere. In constant battle with their own drug habit, Sam and Michelle sympathized. And they needed muscle to safeguard their dream. B. T. figured that he wouldn't get the job, if in fact there was one, because it was rare that anybody hired him these days; he was too smart for the few wage-slave jobs that were available in Twentynine Palms and employers were suspicious if he asked for employment. Generally, he made a preemptive strike and let prospective employers off the hook by telling them he could see why they wouldn't want to hire him. The opportunity that appeared to present itself at the Silver Quarter was different, a job with his name on it. He suggested to Sam that if he was wondering about him, he should check with the kids. Sam declined; anyone who would come in and say what B. T. had just said was a brave man and deserved a job. He gave him a green apron with pockets for making change and put the new recruit to work right away. B. T. tied the hip-length cloth around his waist, a little embarrassed at this goofy addition to his freshly laundered jeans and whiter-than-white T-shirt—he may have been busted out of the Corps on a DUI, but ever the Marine, he could always muster a clothing inspection. One of the kids caught his eye as Sam gave him a supply of coins and he shrugged off his new image—what're you gonna do? "Hey, B. T.," the kid said, "got change for a buck?" B. T. gave him a look and they had a laugh: ice broken, life goes on.

It was true that B. T. had a following. His pad west of downtown on Lupine and then in town on Cholla—closer to the Silver Quarter because his driver's license had been revoked and he wanted to walk to work—was a haven for runaways, kids thrown out of their homes, kids whose parents abused them, anyone who wanted to get away from anything, and there was a lot of that in the desert. The kids felt safe with the big Marine; even when drunk, he'd put up a lethal defense of the defenseless, and if he was planning to be gone for a few days, on a self-prescribed bender, he'd leave his door unlocked and put out the word that all were welcome. B. T.'s sparsely furnished one-bedroom flat was not just safe, it was well ordered. His possessions were neatly arranged, military-style—books stacked carefully on top of a rarely used stove; compact discs laid alphabetically on top of the boom box; the bed made with hospital corners; sleeping pillows carefully covered with Ren & Stimpy T-shirts because the one thing he had not yet purchased since his wife left him after he was kicked out of the Corps was pillowcases. The cupboards contained a few staples—cans of Dinty Moore stew, a couple of boxes of spaghetti, a bottle of Yukon Jack—and were spotless. The refrigerator generally housed a twelve-pack or two of Bud and maybe a T-bone steak. All in all, the pad could be said to follow the instruction given to Marines who found themselves prisoners behind enemy lines: "adapt and adjust."

Among regulars at the Silver Quarter were certain members of the LBG. The Silver Quarter had opened the summer after Mandi and Rosie were killed, and Lydia Flores and her sister Sandy generally stopped by after summer school, instead of going home, hooking up with some of the Crips who passed time in front of the violence-filled video screens that popped every few seconds with cartoony explosions, a weird echo of the armaments that were housed in the world's biggest Marine base a couple of miles up the street. Sometimes Mandi's brother Jason would come in, sit on a bench behind the big window that looked out onto Adobe Road, and just stare. B. T. always asked him what was up and he always said, "Nothing." Everybody knew that Jason hadn't cried since Mandi was killed and they wondered if one day he was just going to lose it.

On a particularly hot afternoon, under a sun without mercy, Mandi's friend Beth DesRosiers walked into the Silver Quarter. She had run away after the murders, staying in various safe havens in California, Arizona, and Nevada, the shabby old homestead shacks that dotted the Mojave and told a tale of collapsed dreams. A few weeks earlier, she had returned home, leaving only to buy drugs and coming back only to pass out. Beth looked pretty raggedy—hair chopped and bleached black-and-white-zebra–style. She wore threadbare jeans and a dirty tank top. She sucked on a Marlboro butt. Sandy, the big and beefy Flores sister, watched Beth as she approached the Doom screen, and elbowed Lydia. They knew that Beth had been one of the last to see Mandi alive and that she had had an argument with Mandi which may have prompted her to head to Rosie's on the night of the killings. Although Beth was bigger than Sandy, and taller, almost six feet, she was high on pot, Thorazine, and probably a number of other things that took her up and down, and she did not notice the Flores girls. "You cunt," Sandy screamed. "Bitch," Lydia added, coming between Beth and the Nintendo game, kneeing her in the stomach. "You could have saved Mandi, you slut," Sandy continued as Lydia pulled her hair. Jason, who had been sitting in his customary position at the window, watched the fight break out, said nothing, and bolted into the heat.

B. T. had broken up many a fight at the Silver Quarter, always among boys, usually Marines and locals. Sandy and Lydia had Beth on the floor quickly and he piled right on, pulling the girls apart even as they continued to punch and kick him, dragging them across the floor, and as Sam held the door open, ejecting them into the street. "LBG rules," Lydia called as Beth hoisted herself atop a video game and caught her breath.

Later that night, after the arcade had closed, B. T. walked toward downtown, carrying his pool cue in its long case. It was League Night in Twentynine Palms and he was looking forward to carrying his team, sponsored by the Break 'n Run, to a victory in a shoot-out with the players from Stumps, a notorious Marine hangout. To cast off the day, he took a meandering route to get to the Virginian, a dive bar on Highway 62, heading east on San Gorgonio, past the Bowladium, through a string of empty lots filled with gravel and dust, and then south on Palo Verde, planning to stop at Andreas on the corner of 62 for chicken-fried steak before the tournament began. As he walked past the little bungalows with the chipped paint, the thick patches of white oleander that fronted some of the yards shivered ever so slightly with the evening breath of the desert. A quick study on matters of survival, he prided himself on knowing that oleander was poisonous, and liked to share his knowledge with newcomers to Twentynine Palms. At 6422, he looked into the driveway, not for any particular reason, but because something, someone had caught his eye. Under the glare of a bare lightbulb outside Rosie's old apartment, he saw Jason, sitting on a crate, staring. Loud music—ZZ Top, as always—came from inside number seven, the pad once again home to those who could not rest. "Hey, what's up?" he called, and approached Jason. Jason shrugged and indicated that nothing was up. B. T. laid down his cue case and crouched in the dirt next to the boy. "What're you doing here?" he said. Jason explained that he came here all the time, was especially drawn to it late at night, even though it was past the town's 10:00 P.M. curfew for minors, a rule he had already been ticketed for violating. "Isn't this where Mandi died?" B. T. asked. Jason nodded yes. B. T. said that after the murders, the cops came into Andreas and questioned him because he was known to wander the streets of Twentynine Palms at night. "Guess they had the wrong Marine," Jason said, and they both laughed. The door to number seven opened and the party inside beckoned; someone invited them in and they declined. "But we'll take a couple of beers," B. T. said. The Buds were tossed, caught, popped, guzzled, earned their reputation: under the glare of the bare lightbulb, moon in this immediate universe of castaways, the court-martialed ex-Marine comforted the teenage boy whose sister was a victim of the Gulf War, the first time since the murders that the Corps had offered comfort to anyone in Mandi's family, unofficial though it was. It would also be the last.

Several months after Mandi and Rosie were murdered, Kevin James got out of jail. He would have entered eleventh grade that year, but he decided not to return to school, using the excuse that he had already missed half the year, so what was the point? The last time he had heard from Mandi was by way of a series of letters she had sent him while he was at the West Valley Correctional Center in Rancho Cucamonga, an hour west of Twentynine Palms, temporary home to a steady parade of desert incorrigibles. She had threatened to kill herself, saying that she couldn't live without Kevin and was waiting for him to get out of jail, hoping that he would change his mind about Janita, his new girlfriend. But then the letters stopped and he was relieved. When he heard that Mandi had been murdered, he blamed himself. He had been planning to go back to his grandfather's house in Two Nine and visit Debie as soon as he was released from jail, but instead he found himself heading for his mother's house in nearby San Bernardino. He hadn't seen his mother in a long time. On the bleak truck runs outside of West Valley, he hitched a ride, as inmates often did, although that wasn't his original plan. A few of his homeys were supposed to have picked him up, but they never showed. The ride dropped him off at the Milliken Avenue exit near the intersection of the 10 Freeway and the 215 north to Barstow or south to San Diego, an ugly confluence of concrete pylons and off- and on-ramps that strongly hinted that it would be a good idea to leave quickly. He walked into the slums of the city that was once a stop on Route 66—what John Steinbeck had called "the mother road"—hoping his mother would be glad to see him, although he knew this was pointless; he figured she'd be too smoked out to care. He had never been to this particular house before, but it didn't matter. No matter what street or town, his mother lived in the same house—run-down, dirty, cheap, duct tape holding up the wallpaper. Here it was, with the dregs of the screen door wide open and the low drone of television voices saying someone had been inside at one time or another. He entered and immediately spotted his mother sprawled on the floor in front of a *Little House on the Prairie* rerun, surrounded by butts and a couple of needles, fading in and out of a nod, way beyond smoked out. He moved her to the couch, rearranged her rumpled housecoat so it covered her like a blanket, said, "Hi, Mom," and left. He locked the door behind him.

On the 10, he continued eastward, hitching a ride through Redlands to Yucaipa, then waiting for a while under the shade of the Yucaipa overpass, then getting another ride through the desolation of Fontana, Colton, and Cabazon (home of local Indian bingo and the Morongo Indian Reservation, perhaps the bleakest reservation in the country), disembarking just past the sign for OTHER DESERT CITIES at the exit ramp for Twentynine Palms and Joshua Tree. He stayed there for a while, thumb out; no ride, few cars. When a cripple in a wheelchair wheeled onto the ramp near him, he knew he had better start walking—the white, handicapped hitchhiker was too much competition for the dark-skinned James, even though the guy looked as if he had been wheeling around the desert for days. Kevin began the walk, traversing Highway 62 up a long grade and into the first town of the higher elevations, Morongo, where the terrain shifts from weird to weirder, where the cacti begin to contort in unexpected postures in order to receive the sharply angled rays of the sun. Kevin stopped at the local package store and bought a Gatorade. It was near sundown and he called his homeys from a pay phone next to the Cactus Mart "All the Cactuses You Can Dig for 39 Cents Each," punching in the number of the phone when he reached their pager. A few minutes later the phone rang. He grabbed it. "Yo, whassup, I'm in Morongo, come and get me," he said. Two hours later, Antoine Lemons and Lebow Pritchett arrived in somebody's new Thunderbird. Kevin got in and they apologized for being late, explaining that they had been waiting to hook up with a couple of Marines and the Marines called three times to say they were on the way but they never showed up, which was why they never made it to West Valley. "Drop me at Mandi's," Kevin said, and they knew he meant the house on Wildcat Drive in Twentynine Palms across the street from the high school. "I ain't seen Debie since she died."

The sun had just about gone down when they pulled into the gravel driveway. Corky bolted from behind the house, planted himself before the car, and barked. "Shit," Lebow said, "you sure you wanna get out here?" Kevin cracked the window and in the fading desert light he spotted Debie pacing up and down on the roof, not paying much attention to the situation in the driveway. "Yo, Debie," he called. She didn't look toward the voice, or respond. Jason emerged from the house, smiled at the sight of Mandi's old boyfriend, and ran to greet him, calling off Corky. "Okay, everything's cool," Kevin told Lebow and Antoine. He got out of the car and Jason hugged him. Corky chased the Thunderbird all the way to the street as it backed out of the driveway. "What's up with Debie?" Kevin asked. Jason said that since Mandi was killed, sometimes she paces the roof, sometimes she chops wood, sometimes, he continued, she just . . . then he shrugged and said, "I don't know."

Kevin James did not go back to his grandfather's house that night. He did not go back the next night. He stayed at Debie's for the next few months, and the two of them reached an unsettling kind of peace. Kevin knew that Debie had not wanted Mandi to "burn coal"; in fact, she had told Kevin many times that she did not approve of Mandi going out with black boys. But Debie knew that Kevin made Mandi feel pretty, had not made fun of her as some of the white boys did because she still had a few extra pounds of baby fat, accepted her as she was. She knew that even though Kevin had knocked up someone else shortly before Mandi was killed, he and Mandi would have gotten together again at some point, would have had a "zebra baby." In fact, they often talked of things that might have been if Mandi were alive. "She wanted to get on *Soul Train,*" Kevin said one winter night as a chill settled across the desert. On this kind of night, the coyotes sounded sharper than they did when it was warm, when the heat baked off under the moon: they echoed each other more keenly, their yip-yapping ricocheted through the cold howl and into the bone of human and perhaps their own flesh alike. Debie took some logs from her pile of wood and built a triangle in the old stone fireplace, wadded up some old *Soap Opera Digests* and stuffed them underneath, then lit the incendiary with a long match. She stretched out before the fire and smoked a cigarette. "Mandi sure could dance," she said. "She got that from me." Kevin pointed out that he had seen Debie dance at the Iron Gate when the bar sponsored Teen Night and he and Mandi used to go and that Debie was nowhere near as good a dancer as her daughter. Debie said that Kevin was right, and offered him a cigarette. "But we were clones," Debie said, and started to cry. The phone rang; Debie grabbed the portable. It was one of Kevin's homeys. Kevin took the call and then flipped on the television. L.A. was going up in flames: the tape of Rodney King being brutalized by the LAPD had just been broadcast and black people were destroying the city that had beaten up their brother. "Come and get me," Kevin said to his buddy on the phone, then, to Debie: "Antoine and Lebow and me're goin' to L.A." Debie sat up right away. "Oh no," she said. "You ain't goin' to no riot." Kevin said that he had to; the time had come. "For what?" Debie asked. "You to go back to jail?" Kevin said that she knew he didn't care about that. "Yeah, well," Debie said. "Mandi did. She tried to reach you and your homeys the night she was killed, remember?" Kevin was hoping that Debie would never bring this up—it was the unspoken truth between them—that Mandi had called the Crips for help on her last night, but everyone was high or in jail, and now here it was. "Oh, so now you blamin' me for

Mandi bein' killed?" Kevin said. "You don't have to, I already blame myself." Debie backed off, saying that she didn't blame him "for nothing," except introducing Mandi to gangbangers some of whom turned out to be more honorable than Marines. There was a knock on the door and honking and Antoine called out for Kevin. "I gots to go," he said, and started to leave. Debie ran to the door and stood in front of it, telling him that if he really loved Mandi, he wouldn't go to Los Angeles to burn it down. "Mandi didn't want no race war," Debie said, then added: "She sure didn't get that from me." Kevin agreed and they shared a begrudging laugh. As Reginald Denny was getting his head kicked in on the corner of Florence and Normandie with the simulcast on television for the world to witness, Lebow honked and Antoine banged again on the door and called for Kevin. "Changed my mind," he called through the door. A few days later, he moved out, stayed at his grandfather's for a few days, then hit the streets and was soon back in jail. Some said it was for driving without a license, others said it was for possession of marijuana. On the outside again a couple of years later, even Kevin himself would not remember what infraction, exactly, had deprived him of his freedom, such as it was.

After Mandi died, her older sister Krisinda began having nightmares. For a while, it was the same one. Mandi's hair was cut short. She was in a wedding gown. Krisinda was her maid of honor, with long hair. As the ceremony was about to take place, Mandi was stabbed and Krisinda was splattered with blood. To change the scenery while she was awake, Kris moved to her own small house in Twentynine Palms, at a slightly higher elevation than Debie's, north of Highway 62, near the first rise of the Queen Mountains on Sunnycrest Lane. She was visited by a different nightmare. A large unidentifiable form came in through her bedroom window and attacked her. She became an insomniac, afraid to sleep. One night, after she had finally faded, she actually was attacked, she said—a Marine broke in, and as he was about to attack her, she screamed and he fled. She called the cops but nothing ever came of it. At sixes and sevens, she joined the Army. Debie did not initially approve of the decision, worried that her only remaining daughter had signed away her life for what, really? An oil field? But Kris convinced Debie that the military was indeed her only escape from Twentynine Palms, and that there was really no other way the family could afford for her to go to college, or any kind of technical school, and learn skills that would increase her value in the job market. Debie finally agreed, and even helped Kris get into the ranks earlier than the usual wait time by writing their congressman and asking him to help her only remaining daughter start a new life. But when it was time to go, Kris was too thin; she had lost so much weight so quickly, twenty pounds since a physical two months earlier, that the Army changed its mind, suspecting that she had an eating disorder. To ease her loneliness and share her rent, she took in a roommate, a cashier at Benton Brothers, the local family-run general store. She went into therapy on the base, taking advantage of her status as the child of a veteran, although she hadn't heard from her father in many years.

On the surface, she seemed to be improving, understanding, she said, how she had been drawn into a series of abusive relationships with men and that it was possible to break the pattern of the ages. To boost her self-confidence, she enrolled in a public-speaking course at Copper Mountain Community College, and unveiled her new skill by way of a speech that she gave inside the belly of the beast, at the Marine base, to a group of spouses of Corpsmen, some of whom knew that Krisinda's sister had been killed by a Marine. It was called "I Remember" and it was about Mandi.

I remember bits and pieces of the past. The first was when I was six and a half years old. Mandi and Jason were running back and forth filling a swimming pool. They collided and they had to be rushed to the hospital. I cried because I didn't understand. Jason got stitches in his hand and she at four lost her front teeth. Then she finally gets her front teeth back and one day while playing baseball in the park she chips them . . . I remember bits and pieces of her life. Like the time she thought she was "one of the guys" and she tried chewing tobacco and it made her sick. I remember the day in Sacramento when we were all playing and she ran out the door and was hit by a car. The car ended up with a dent, she only ended up with a sore arm and a little road burn. She came home laughing and happy to have hugging from the hospital monkeys. I remember being in a lot of trouble and my mom hitting me, and it was she that came to save me from my mom's wrath. I remember the day that she announced that she would not have any kids but adopt and raise monkeys instead. I remember when she was on the Park and Rec. football team. She was the only girl. I remember her birthday at Chuck E. Cheese, how she was fascinated with the singing animals. I remember her first crush and I remember the time she came to me because she thought she was pregnant. Although it turned out to be a false alarm, it brought us closer. I remember she was there for me when I got my heart broken . . . And I remember her love for people, how easily she made friends. I remember her love for dance music and how much she loved to dance. I also remember our fights and how somehow or other we always made up. I remember her eyes, her hair, some-

> *times even her voice. Most times I remember happy times like fighting over the bathroom or keeping secrets. Oh, I remember bad times also, but the good times far outweigh the bad, and that's the way I like it. She was a wonderful, happy person, she had her flaws, but every person does. She gave a lot of herself without expecting too much in return. She was a rare breed of character—funny, happy, spontaneous, and much more . . . She is my sister, someone whom no matter what, I'll always remember. She is just away!*

Her speech was what she thought everyone wanted to hear, the kind of hyper-chirpy have-a-nice-day varnish on tragedy that has come to pass as grief. But several months later, after her boyfriend had broken up with her, Krisinda took a handful of sleeping pills in front of her roommate and passed out. Her roommate called an ambulance and Kris was taken to the High-Desert Medical Center, where her stomach was pumped out. When she regained consciousness, she did not remember how she had come to land in the ICU. She explained to her mother that she just wanted to sleep.

When she was released from the hospital, Kris moved back in with Debie. Debie cooked her favorite dishes—biscuits and gravy, roast chicken, potatoes O'Brien—and over a period of weeks, Kris stopped having the nightmares and gained back some weight, although not as much as Debie hoped. It was found that she now had a bleeding ulcer. But she felt strong enough to move back to her own house and ready to look for a new job. In this town of endless job turnover, she quickly found one, at the video concession inside Downtown Liquor and Deli on Highway 62, at the heart of Twentynine Palms, a few blocks away from the murder scene.

One night, Chad Yerkey came in to rent movies. He was tall,

lanky, handsome in a guileless, Midwestern kind of way that always seemed to serve as the projection screen for the hopes and dreams of girls who had not yet formed strong identities and were unable to see anything other than the answer to the vague yearnings they thought they were supposed to have on the face of the man who was supposed to fulfill them. Kris could tell from the buzz cut that Chad was also a Marine, a guy with a steady paycheck and periodic upgrades if he performed well. She agreed to a date when he asked her out, and then there were more dates, and soon, not long after Mandi's murder, to Debie's dismay, her remaining daughter was in love with a Marine.

Chad was sent to Somalia and Debie hoped that Kris would lose interest. But her heart responded to his absence, and she joined one of the support groups in town for friends and spouses of Marines who had been sent to that particular "conflict zone." They wrote letters. They baked cookies. They sent sexy photos. When Chad returned, he and Kris resumed their relationship, often double-dating with Chad's roommate from the Corps, a guy from Jacksonville, Florida, and his girlfriend, Tammy Watson. Kris had told Chad about Mandi's murder, warning him about her nightmares which periodically recurred, and that a Marine named Valentine Underwood was being held for the crime, awaiting trial, the scheduling of which had undergone delay after delay. Over pizza at Rocky's one night, Kris began talking about the murder of Mandi and Rosie, and how another six-month delay had just been granted and when would this guy Valentine Underwood come to trial? The generally loquacious Tammy suddenly clammed up, as the others continued the conversation about how money greases the judicial system, laughing about that old gag that everyone was created equal, not me, no way, jeez what do they take me for, some kind of idiot? Kris asked Tammy why she was just staring at her food—she was the one who wanted pepperoni pizza and now she wasn't touching it. "I have to tell you something," she began. It was difficult to find the words. Here was the sister of a girl who, Tammy believed, had been murdered because of Tammy's own cowardice and her trust in her father and the Marine Corps. Here she was, beginning a new relationship for the first time since she had been raped, and she had not wanted anyone—particularly a new beau—to know about the incident. And here she was, with two Marines, who would now hear of a terrible failure of their military family.

"Kris, you won't like this," Tammy said. "It's all right," Kris said, "go ahead." Tammy added that none of them would like what she was about to say and then she blurted it out in one endless breath: "Valentine Underwood raped me before he stabbed Mandi and Rosie and I thought he was supposed to be restricted to the base on the night of the murders because I told my father who's the highest-ranking black NCO in the Corps and my father said he would take care of it so I dropped the charges." She was in tears now, and Kris, sitting across from her, quickly exchanged seats with Tammy's boyfriend and held Tammy as she sobbed. Kris told Tammy that it wasn't her fault. Over and over, Tammy said that it was. There was nothing for the guys to do, so they got up and left, bereft of gestures or words in this emergency that did not require a physical rescue. Chad laid a twenty-dollar bill on the table and said he'd call Kris later from the barracks.

Now that her secret was out, Tammy found a great deal of comfort in talking to Kris. Kris, also violated, although indirectly, by Valentine Underwood, was more than happy to listen; the two formed a quick and deep bond. Between double dates with their boyfriends, they would talk for hours on the phone. Tammy, just a year older than Mandi when she died, became kind of a new younger sister to Kris. One of the things they often talked about was whether Tammy should go public with her account of being assaulted by Underwood. Tammy was reluctant, as it raised too many questions about what her father might or might not have been able to accomplish after Tammy had told him about the rape. And he was going to retire soon with a perfect record. Why not just let the whole thing slide?

Kris sympathized but had told Debie about Tammy's night with Underwood. Debie had immediately embarked on a research project that would consume her life, reading up on how to file a wrongful death suit against the government, figuring that if the Marines knew they had a rapist in the ranks, and they permitted him to range into town, and one night while doing so, the alleged rapist had murdered her daughter, wasn't the government liable? To Debie, the answer seemed clear, and she tried to retain a lawyer, calling the high-profile plaintiffs' attorneys in the area, and even a few stars in Los Angeles. "You don't know me," she would begin. "I'm just a worker-bee bartender and my daughter was killed by a Marine who should have been locked up and I'm wondering if you'd take my case." No one was interested, and she resorted to the yellow pages, dialing 1-800-boohoo, as she called the numbers, contacting lawyers who said they specialized in suing the federal government. She called at least six or seven, and was turned down as many times, each time being told that the case was just too difficult to prove. Finally, she went to a stationery store and picked up the lengthy do-it-yourself sue-the-government forms. But there was a problem: the case depended on Tammy Watson, and even after many discussions with Kris, Tammy still did not want to come forward, especially now that she had heard there might be a lawsuit, which meant, among various unbearable scenarios, that her father would have to testify against his beloved Marine Corps, telling the world that Tammy had told him about the rape and that he had told her the Corps would look into the matter. But Kris convinced Tammy to meet Debie, and then make her final decision after that. Tammy agreed. When Kris and Debie arrived at Tammy's house for the arranged meeting, Tammy wasn't home. Nor was she home at the next designated time, or the next. Finally, weeks after she had flaked on the meetings, she phoned Debie and said that she'd like to talk.

Once again, Debie and her family found itself wandering through the desert, deeper this time, heading for another oasis. This journey involved what people in the Mojave referred to as a river run, a quick trip made for a variety of reasons to take in, in a variety of ways, the healing waters of the beleaguered Colorado River. Kris and Chad were getting married in Laughlin, Nevada, a casino town well northeast of Twentynine Palms, on the banks of the pulmonary vein that runs right through America's plains and canyons of scorched earth. On an October afternoon, Debie began the long drive, headed east on Highway 62, past endless Mojave playas and ancient dry lake beds where the saber-toothed tiger had given up the ghost twenty thousand years ago and new incarnations of flesh and bone had themselves come and gone, reclaimed by the advancing white sands. Just past the Turtle Mountains, at the intersection of Route 95 and the 62 in the town of Vidal Junction, Debie refueled and picked up a bottle of Southern Comfort in the package store across the street from the gas station. She had wanted to bring Kris a bottle of good champagne to crack open at the wedding reception, but was not able to find an affordable one earlier in the day, and the only champagne in Vidal Junction was André, so she decided on the Southern Comfort. Debie continued north on the 95, and, in the way that renders time pointless in the desert, paralleling the Chemehuevi Valley Indian Reservation and—due east of the reservation—Lake Havasu City, a party and retirement town which had sprung up around Lake

Havasu and had recently become home to London Bridge, the very walkway itself, imported from America's motherland stone by stone and reconstructed in Arizona to attract visitors. In Twentynine Palms, some people considered it good luck to go to London Bridge and kiss under its arch (it forded no river, not even a tributary of the nearby Colorado; it was just too small). Debie didn't know if Kris and Chad had stopped at the bridge, on their way to Laughlin earlier in the day, or ever, but she remarked that they would need a lot of luck: the cards in this particular deck were stacked against both players.

Evening had set in and the approach to Needles lowered to an elevation of 480 feet as the 95 crossed Interstate 40 at Needles, site of the army's Fort Mohave during the Indian Wars of the 1800s. A quarter moon refracted off a welcome glimpse of water in the desert; here, the Colorado River poured into the northern tip of Lake Havasu and the 95 carried travelers across the flow, through the Fort Mohave Indian Reservation, and into Bullhead City, Arizona, home to blackjack dealers, waitresses, residents of floor shows, and others who worked across the river in Laughlin. A short distance to the east was the town of Oatman, Arizona, and then Kingman—where Timothy McVeigh had found sanctuary and wide-open space and the others who also dwelled within and nearby echoed his bet, his belief, that it was a free country and he could do whatever he wanted to do, and no one was here to stop him, or any of them—that's what the teachers had promised.

Debie checked into the Best Western in Bullhead City, quickly changed into her mother-of-the-bride outfit, red leather pants and jacket, and drove to Laughlin across the Davis Dam, a vast barrier whose grim beauty as an engineering feat went unnoticed, it was simply the conveyor to or from casinos. Kris had given Debie directions to the chapel, but it was difficult to locate in the dark. After driving up and down the main strip several times, she spotted it, perhaps having unwittingly missed it because she had hoped that Kris would not begin her life as a grown-up as she had, with a quickie wedding. She parked in the adjacent lot and rushed in, the last to join the small wedding party, which consisted of Jason, who had arrived earlier with Kris; Chad's parents, who had driven down from their small hometown of Coal Valley, Illinois; Uncle Mitch—not really anybody's uncle but a close friend of Debie's family who had come from Twentynine Palms along with his wife to give the bride away; Kris's friend Kim and Tammy Watson, both maids of honor, assuming the role that would have gone to Mandi. A husband-and-wife team named Alice and Ted presided at the ritual. They appeared to be as old as the desert, and were just as relentlessly sunny in the face of so much nothing, such a vast parade of unions that masqueraded as the fruit of dreams but were born really of desperation and haunt. So glad you could come to the Yerkey wedding, they said to each guest upon entry, leaning in close to offer a hearty handshake and a fresh smile, and then offering their own brand of confession, seeking a kind of redemption in their little rented chapel in the Mojave, the wife would say, "We don't smell, do we?" and the husband would add cheerfully, "We brush our teeth with tea-tree oil. We use it on our skin. It prevents age."

With everyone seated in folding chairs, the wife turned on the recording of "Here Comes the Bride," a baroque harpsichord rendition that sounded a bit tinny. Chad, in full Marine dress, waited at the front of the chapel. Kris, pretty as any bride in her new gown, entered on the arm of Uncle Mitch, sporting his motorcycle leathers. Debie grasped Jason's hand tightly as Kris finished her walk down the short aisle, too tight for Jason's comfort, as Kris let go of Uncle Mitch's arm, leaving the last protection that Debie could muster for her remaining daughter and now facing Chad, the latest Corpsman to enter and affect the family. The music faded and the ceremony began. Neither Kris nor Chad had requested any particular reading or religious citation, so Ted recited his standard, an Apache homily that mentioned cloudy skies and how harmony can clear them and it was quickly finished and bride and groom were pronounced man and wife and kissed quickly. Debie handed Alice the bottle of Southern Comfort and she quickly opened it and poured some into the little plastic champagne glasses that the chapel provided and then she quickly ushered Debie to the podium, where she announced that the bride's mother wanted to make a toast. Debie raised her glass and, ignoring an earlier promise to Kris to maintain decorum, told a bawdy joke about a bride and groom on their wedding night. Jason, Tammy, Mitch, and his wife laughed. Chad's parents did not. The bride and groom smiled politely. Alice checked her watch and gestured to Ted. Congratulations, he said to Chad's parents, approaching them and leaning in to offer another hearty handshake. Congratulations to you, he told Jason. What a lovely ceremony. The bride looks beautiful. I'm sure they'll be very happy. Thank you so much for booking with our chapel. Please tell your friends about us. A new name flashed across the chapel's message board now—WELCOME, MONTEITH PARTY—and Ted and Alice ushered the Yerkey party out, waving their wrists up and down in tandem to express good-bye and then heartily pumping the new round of incoming, clammy hands. Hope my breath is okay, Alice said. We use tea-tree oil. It's the secret of a long marriage.

Chad and Kris honeymooned at Harrah's in Laughlin. They ordered surf and turf in their room. They were comped all over the casinos with free drinks, as everyone was, especially newlyweds and members of the military. A few months later, Chad completed his tour in the Marines. He returned to his home deep in the guts of the Midwest, Coal Valley, with his California girl Krisinda, and her two pet ferrets, her new pit bull, and her parrot. They moved in with his parents, into the house where he had grown up, as had his father before him, and his grandfather before his father. He got a job as a diesel mechanic. Kris found employment as a technician in a veterinarian's office. As mileage goes, her wish had been fulfilled: she was far away from Twentynine Palms, and as the first blizzard of the year swept across southern Illinois, the first snow that Kris had ever seen, shutting down Coal Valley and forcing families to remain in their homes and talk to each other, came the first notion that perhaps she had married too hastily, although these were not the words that Kris would attach to this notion, she simply did not feel right; she felt the way other, well-to-do girls felt on their first night in a strange dormitory when they started college far away from home, worse, perhaps, for the other girls had many choices. She excused herself from her circle of in-laws and called her mother.

Some old strays had moved back in with Debie—the teenage speed freak and her young son. Mandi had been the boy's regular baby-sitter and he had not been the same since Mandi died. Debie was trying to wean the girl away from speed, a difficult task in the land of cheap crank from home cookeries in hidden dens that dotted the Mojave. Sometimes the girl would disappear for days, but she would always leave her son. Debie would take care of the seven-year-old while Jason was in school and then Jason would watch him while Debie worked. Sometimes Debie would bring both Jason and the boy with her to the Oasis so she could keep an eye on Jason, who had crawled into a hole deep inside of himself since Mandi died, talking to few, hardly ever going to school, connecting with others only if the occasion had something to do with his sister. When he was with the boy, Jason would read to him the way Mandi used to. He wasn't a very good reader, but he liked taking over Mandi's role. He read whatever was at hand, often the liner notes from whatever CD might be handy.

In memory of Mandi, Twentynine Palms High School had planted a tree in the central courtyard. It was a tree that had no name—no one in the Lunch Box Gang knew what kind it was; nor did the school groundskeeper. It was like a lot of things in the desert. One day, it just showed up, no fanfare, no witnesses, someone saw it and named it—in this case, it was the Mandi Tree—and the fact that it was in town was passed on amid the news of the day: "By the way, they planted a Mandi Tree, did you hear?" "Yeah, Mandi's got a tree, isn't it cool?" "Wanna go see the Mandi Tree?"

Sometimes Jason would start walking the streets of Twentynine Palms with the little boy. Most of the time they just walked. Jason did not want to show him the apartment where the murders happened. So they would walk around it if they were heading in that direction. One day they found themselves at the Mandi Tree, which was a long walk from the Oasis. The Mandi Tree was kind of scrawny and withdrawn, self-effacing, almost; it wasn't particularly leafy and provided little shade in the desert sun or cover in the monsoons of the late summer and early fall. It was the first time Jason had been to the tree; he had heard from Mandi's friends that it had been planted but he hadn't been able to bring himself to the grounds because he thought it would be too sad. But he had finally conjured an image; although he would not have put it this way because the tree he pictured gave him a feeling, not words; he had been expecting a big tree, with ancient branches and lots of leaves, the kind of tree that you saw in movies about places where they had trees that people went to because they reminded you of someone, and when he saw the scrawny, fledgling Mandi Tree, he tried to make the best of it, because he was with his young friend and he had told him they were going to see a really cool tree. The boy held Jason's hand

and Jason said, "Here it is." They looked at it for a while, then they walked back into downtown Twentynine Palms. "We just saw the Mandi Tree," the boy told everyone in the Oasis, and Jason shrugged and said, yeah, it was okay, you should go see it, his marrow aching with the truth—that it was a nice try, but like a lot of things in the Mojave, was already suffering from want. Debie suggested that Jason take the little boy home for a nap, knowing that it was really Jason who needed to rest.

Generations of snakes use the same lair, crawling back and crawling back and crawling back again, all the way home. So, too, does every town have a piece of real estate that attracts the same population, and as such, it is jinxed: the stadium where the home team always loses, no matter what sport the team plays and how often its members are replaced, the vacant space that no one wants to rent even though dozens of people walk past it every day and it's a great place for a store. In Twentynine Palms, the location that would bring bad luck would seem to be apartment number seven at 6422 Palo Verde Street, scene of the murders of Mandi Scott and Rosalie Ortega.

Rosie's mother had been correct to suspect that here was a stage for foul play: shortly before Rosie had moved in, the previous tenant had been raped at knifepoint by a Marine. After the murders, Rosie's mother and her sister packed up Rosie's things, scrubbing down the apartment even though they knew all signs of violence would be removed before the apartment was rented again; they couldn't just pack up Rosie's things and leave her blood all over the place, so they scrubbed and scrubbed and wrung out the rags in the bathtub, already splattered with Rosie's blood, and they scrubbed until it was as clean as they could get it and then they turned off the lights and got in their Honda and left. The apartment in the sad little complex was quickly refurbished with a new paint job, new fixtures, new shag carpet. It had no trouble attracting tenants, haven that it was for people who were going under, just gaining a toehold, or holding on for life. One way or another, it sent them on their way, always ready, like the desert which produced it, to provide sanctuary for all who entered and warmed themselves at its fire, always, it seemed, ready to pounce when those who took up residence let down their guard. It was spring, a starlit evening filled with the skittering of fledgling critters, the kind of night that desires its inhabitants to hook up, that twinkles even more brightly at the energy refracted when all who dwell within it exchange phone numbers, DNA, a look. Number seven at 6422 Palo Verde was host to another all-night party. It is not known exactly who was the official tenant at the time, but the guest list included Marines and young girls. As the local bars closed for the night, late arrivals began to stumble in, including a Marine private. He met two sixteen-year-old girls and they all polished off a six-pack. Like almost-sixteen-year-old Mandi before them, the girls went to sleep in the bedroom. The Marine approached one, pulled her pants down, and inserted a beer bottle. She screamed and woke her friend. The girls jumped on the intruder, hitting and cutting him with the bottle, then ran to a pay phone and called the police. Still inside the apartment, the Marine was arrested for rape, bruised and bleeding, a shadow player in a piece that had been playing here forever, it seemed, until someone could come up with a different ending, let some light into this dark little desert pad and the lives it contained.

Two years passed. Mandi would have graduated from high school, hopped in the Camaro, and headed out of town, for the future. But the Camaro was parked in Debie's driveway. Occasionally, someone would drive it but never very far, just down to the store or into the desert for a cruise. Some members of the Lunch Box Gang invited Debie to attend their high-school graduation ceremony. She had said she would, and was even planning to make a speech about Mandi and what she meant to the students of Twentynine Palms High School, but several days before the event she was overtaken by a crying jag so she skipped it. Instead, she planned an event that would mark the milestone in a way that was more appropriate to Mandi. She planned a party. Not a typical Twentynine Palms party—though of course it would be that—but a fund-raising party. Debie was starting the Mandi Scott Scholarship Fund. It was not the kind of scholarship that the school would have sanctioned, for it was not based on excellence or achievement of particular note in any particular area, even attendance; it promoted nothing conventional and, on its face, nothing specific. What Debie hoped to do was raise a thousand dollars to help an average girl get out of Twentynine Palms. The suitable applicant would have a C average and a plan to "go somewhere and better herself," as Debie said on the flyers she posted around town. She did not know that David Letterman had once offered the same thing on his show—a thousand-dollar scholarship to a student who fared as poorly as he did in high school—as a joke of course—nor was she kidding, although there were some local establishmentarians who found the idea laughable. But considering the local demographics (which Debie had done simply by watching Mandi and her friends), the Mandi Scott Scholarship Fund was just what the desert needed.

The statistics for the children of San Bernardino County were a sad and relentless testament to the ill winds that had carried many of them into the desert. Near the time of Mandi's death, the median family income was $36,000, compared with $42,700 for the state in general. The county's unemployment rate (8.2 percent) was higher than both California's (7.7 percent) and the country's (6.5 percent). Among the fifty-eight counties in California, it had the fourth-highest child population, with 31 percent of all residents under the age of eighteen. One in five children was living in extreme poverty (almost twice as many as the national average), with "extreme poverty" defined as a three-member family with an annual income of less than $8,328. One in ten children was at risk of abuse or neglect. Among unmarried girls who were fifteen to nineteen, more of them were having babies than girls the same age in the country in general. Hundreds of newborns had been exposed to alcohol, drugs, or both. The infant mortality rate was higher than it was for infants elsewhere in California. Among children living in poverty, 82 percent did not receive necessary dental care. On their eighth-grade reading achievement scores, a required statewide test with a perfect score of 500, students scored an average of 248, below the state average of 257. Twenty percent of high-school students dropped out before graduating, just as they did in the rest of the state. In Twentynine Palms itself, at the time of the Mandi Scott Scholarship Fund party, there were nineteen students who were pregnant and seven teenage mothers.

The party was at the Oasis. The theme of the party was purple, Mandi's favorite color. The paper plates were purple, the napkins were purple, Debie wore purple jeans and purple high heels. The Jagermeister girls wore purple bikinis. Word of the celebration spread quickly across the desert, as word of parties always did. Starting at seven in the evening, pedestrians and cars from points east, north, west, and south filed into the bar. Some of the guests knew the party was to raise money for Mandi Scott, and they dropped what money they could afford—quarters, dollars, food stamps—into the big jar Debie had placed on a table at the bar's entrance. Others just showed up because they had heard there was a party, and that the local band, Velvet Hammer, was playing. When told the party was a fund-raiser, most of them would just shrug and head to the dance floor or bar, saying that they had just gotten laid off, or hadn't worked in eight years, but when further apprised of the nature of the fund-raiser—"my daughter was murdered and I want to help an average girl like Mandi get out of town," Debie would say—they nodded knowingly, dug into their pockets, and gave. In exchange, they were provided with the best food the desert had to offer. Mandi's friends had cooked and donated their specialties: the Samoans had roasted a pig and brought slabs of pork; Lydia and Sandy Flores had baked cupcakes decorated with M&M's in honor of Mandi; some of the parents whose kids Mandi used to take care of brought baked ziti and macaroni and cheese and apple pie. The Oasis supplied free taquitos and cheese sticks and carrots and chips and other bar snacks, and Budweiser, the area's biggest distributor, donated the beer. While Velvet Hammer cranked rock-and-roll covers all night long, and the joint overflowed with the many and varied members of Mandi's magic circle, the scholarship jar filled with small denominations of coin and cash. "All I need is one thousand dollars," Debie would tell arrivals, "come on, gimme a buck, a quarter, I know you got it, and if you don't, give me your lottery ticket, empty your pockets, it's a good cause, and we'll get you loaded." Sometime during the night, a barfly known as Mojave Bob arrived, having ambled over from the Virginian a couple of blocks west on Highway 62 and on his way to the Josh Lounge for his customary nightcap. "It's official," Debie called. "Mojave Bob is here. Got a donation for my daughter?" Bob removed a cigarette tucked behind his ear and dropped it into the jug. Debie thanked him and said that Mandi liked menthols. "Good taste," Bob said. "Hey, everybody, look what Mojave Bob gave," Debie said. "His last coffin nail!" Everybody laughed a cigarette laugh and a couple of people called out hey, Bob and then the party kicked into high gear. Bob picked up his free beer and greeted people in his customary fashion, with bar calling cards. He handed a pink one to a drunk lady in too-tight jeans and tube top. "Whenever I Look At You," it said, "I Get An Overwhelming Desire"—and then, in small print—"to be lonesome." The drunk lady cackled and said it was funny, even though he had given her the same card last night.

In between making the rounds, Debie pored over the twenty-one letters of scholarship application. She had asked applicants not to sign their names because she wanted to make a decision based on the merits of the letter, not on personal knowledge. Each was compelling in its own way, presenting a case for the bettering of the self. "I want to get out of town so I can find my father and get him to come back and take care of my brother," said one application. "If I win the thousand dollars," said another, "I will go down the hill and never come back. I don't know where I'll go, but it's a start." "I want to give the money to my mother so she can buy a new transmission and we can leave," said one girl. Debie made her decision as the band played its final song, "Proud Mary," always a crowd pleaser, although Debie could never understand how anyone could possibly dance to it—in fact, it was a private joke she and Mandi shared, how funny this song made people look when they tried to dance to it and how come they all thought they looked so cool?

Debie and Jason lugged the big bottle with the donation money to the stage. Debie said it felt like a thousand dollars in quarters. Velvet Hammer played a drumroll and Debie unfurled the winning application, written on lined three-ring notebook paper with the words overflowing into the margins. "And the winner of the Mandi Scott Scholarship Fund," Debie said, Academy Award acceptance–style, "is . . . will the person who wants to be a juvenile probation officer please step up?" Seventeen-year-old Mimi Quinones screamed, ran to the stage, and hugged

Debie. A couple of the Crips and Bloods and Samoans hissed at mention of "probation officer," but mostly the crowd cheered. Debie explained that although every applicant was a good one, she picked Mimi because "she said she wants to return to Twentynine Palms when she finishes her training, and we need all the help we can get." A couple of the Crips said shit, we don't need no help, but when Debie handed Mimi a check for one thousand dollars, they quieted down and joined the crowd in another round of raucous approval. "Hope you don't mind if it's post-dated," Debie said. "Gotta bring this jar to the bank." As it turned out, Mimi explained to everyone, she was an acquaintance of Mandi's, a little overweight like Mandi, the occasional object of teasing. Once, in choir practice, someone called Mimi "tons of fun" and Mandi had told them to fuck off. "Yeah, baby," Mojave Bob called out, always handy with a joke, five or six pitchers of beer beyond his torqued-out arrival condition. "Fuck off! Fuck me! Eat me! I got something you can fuck!" Debie told Bob that she loved him to death, but the party was over.

The statute of limitations for filing a wrongful death claim against the federal government was about to pass. Debie had picked up the standard form 95 for filing a claim for damage, injury, or death soon after Kris had introduced her to Tammy Watson and she had learned of Tammy's rape, and had filled in most of the blanks—name and address of claimant, type of employment (military or civilian); marital status, date and day of accident, and description of accident. But she had been unable to fill in the box marked "amount of claim (in dollars)" under the wrongful death category. Was her daughter worth one million dollars? A hundred million? How could she put a price on Mandi? She lit up a joint and paced her kitchen. The forms were on the table and she had a little more than an hour to complete them, get them notarized, and then postmarked before the day's legal end.

She sat down, finished off the joint, and thumbed through a little red phone book, looking for the number for her old friend Kat. She figured Kat would be able to help her put a price on Mandi, and would also sympathize about how awful it was to have to do it. Kat knew Mandi, knew how she was Debie's "clone," as Debie always said, knew how close mother and daughter were. Kat's numbers were crossed out and rewritten so many times—more than a lot of the other people in Debie's phone book who moved a lot, or simply changed phone numbers all the time—that Debie had to narrrow her eyes and squint to make out the latest number. She dialed and tapped on the table as she waited for Kat's phone down at the biker pad in San Diego to ring. But it didn't; in fact, what Debie feared would happen, happened: the recording saying that the number was temporarily out of service came on. Forty-five minutes to go before the post office closed. Debie called her friend Mary DesRosiers, at whose house Mandi had argued with her daughter Beth before leaving for Rosie's on the night of the murders. Someone in Mary's family was always in trouble; she'd know the ins and outs of legal forms. Mary's son Ted answered. Mary wasn't there, he said. No, she wouldn't be back soon. Not today anyway. They had her locked up again, he said, in the mental hospital. She freaked out again, he said. His father had punched her out and she lost it. Debie lit up and put down the phone, started to dial someone else, then decided the hell with it. She got up and paced at the shelf with the urn that had Mandi's ashes in it. Man, man, man, she muttered to herself, using her nickname for her daughter. One million? Nah, too cheap. What do you think? She picked up the urn. "I'm sorry, baby," she said. "You don't know . . ." She cradled the urn, rocking Mandi back and forth. After a while, she returned the urn to the shelf, sat down at the kitchen table, and filled in the last blank: five million dollars. "Let's do it," she said to Corky, and he followed her into the pickup and they drove into town.

Word of the filing spread quickly. The notary public evidently told his wife who worked on the base and she told someone at the PX and then it seemed like every Marine at the knave level knew that a popular local bartender had sued the Corps. Since Debie herself had also told both friends and customers, some Marines even knew that the bartender was Debie and that she had a son named Jason and that Jason attended school at Twentynine Palms High. A few of these boots asked their sons, who also attended school at Twentynine Palms High, if they knew the boy who was the son of the civilian who was calling the Corps on the carpet. Yes, they knew Jason, they said. Over time, they waged war against Jason. It began like all wars, with name-calling, some of which had to do with Mandi and her history with boys. Soon it escalated to pushing and shoving. One day after football practice, the coach, a popular NCO who was also the coach of the Marine basketball team, took Jason into the locker room and warned him to stop talking about how Underwood had come to the game that day at Norton with a cut on his hand and had not been able to play. When Debie learned of the episode, she paid a visit to the school principal, who in turn called the cops. Debie told the cops that her son was being harassed because a Marine had killed her daughter and her son had dared to talk about the circumstances. The principal apologized to Debie and Jason, but Debie feared for Jason's well-being; he could defend himself physically, she figured, but his depression since Mandi had died grew deeper by the day—she feared he might harm himself. Soon she had him transferred to Monument High, the alternative school in Twentynine Palms. But Jason withdrew even more deeply into himself, so Debie pulled him out of school and sent him to live with friends in Bullhead City. They ran a tattoo parlor and they gave Jason a job in the shop. In the land where new families instantly manifest and then vanish and then manifest again like the rare desert pupfish after the summer monsoons, Jason immediately felt at home. The guy who ran the place was just a few years older than Jason and became an instant buddy, showing him around the area, pointing out which bars had the best pool tables and where to get the cheapest cigarettes. He was married to a seventeen-year-old girl who reminded Jason of Mandi, except she had two babies. Sometimes she and Jason would sit in the sun on the curb with her babies outside the tattoo parlor and Jason would tell her little things about Mandi. Like how Mandi would use her baby-sitting money and take him for pizza. Or how Mandi had once spent hours and hours teaching him the Roger Rabbit, a dance she had picked up right away. The tattoo needles buzzed quietly as they talked, marking the time; silence for a few hours meant it was time to close up. Since his arrival, Jason himself had gotten a couple of tattoos, starting with his right forearm: the very first emblem was a big cross with a heart at its center, wrapped with a ribbon that said SIS and underneath it the notation IN MEMORY OF MANDI 8/4/75–8/3/91. There wasn't much to do in Bullhead City, especially if you didn't have wheels and couldn't get to Laughlin, so during his off-hours Jason began to add tattoos. None of them were sentimental. At the suggestion of the guys in the shop, next, on the outside of his left leg, came a pit bull with a choke chain and fangs bared; it said MAMA'S BOY. Then came the white power signature—lightning bolts—on his left wrist. Then, on his left thigh, over a three-week period because it was a big tattoo and it was applied to a fleshy area which meant that the procedure had to be done slowly because tattoos applied to fleshy areas cause a lot of pain, came a rooster with a noose around its floppy, broken neck. Jason often spoke of his identification with it. I'm a dead duck, he would say, and smile and shrug. That's me. Over time, the guys at the shop would show Jason to prospective customers, proud of their work. "Guess what we're gonna put here," they would say, pointing to the empty canvas of Jason's back, and then, "Hey, Jason, show 'em your stomach." The young boy had become the shop guinea pig, and for a while, not knowing what else to do after the death of his sister, and reveling in the attention, he played the part. When asked—and he would never volunteer, for he was not a braggart—he would roll up the sleeve of his shirt so a stranger could see the Grim Reaper stalking the tendons of his right arm, and then, he would continue the revelation, and on his right shoulder there would appear a jailhouse standard—two skull faces, one smiling, one crying. "Smile now, cry later," one of the guys in the shop would say, shrugging as he uttered the tattoo's code, the Mojave's echo of the face of comedy and tragedy so familiar to those who are well schooled in the arts and take the image as a metaphor, not a way of life, shrugging in defense as if the desert, the entire universe, required a response, "what else are you gonna do? Hey, J, show him the little guy." Then Jason would peel back the left sleeve, underneath which peered a small, crudely inked Yosemite Sam character brandishing a shotgun next to a street sign marked OUTLAW. By the time Jason returned to Twentynine Palms, he had thirteen tattoos, including a giant gargoyle on his back and an unfinished skull on his belly. It was unfinished because the pushing of all that blue ink into such a tender region had caused Jason too much pain and he told the guys that even though they had done the head, the eyes, and the mouth, he didn't want the nose or the chin, he didn't want any more tattoos. Yet the protest was too late: the little boy who had once trundled through desert playgrounds with his favorite sister had become a walking canvas of death. Debie would have paid the guys at the tattoo shop a visit but her car needed brakes. So she called them to tell them she was not happy about the condition in which her son had been returned. There came a sickening although familiar recording: the phone had been disconnected. Oh well, she thought. Smile now, cry later. And you will cry, motherfuckers.

Nine days before Christmas 1997, Gary Bailey presented his closing argument. In coming to a verdict, jurors must ponder "who could have done such a terrible and vicious and brutal murder; how could someone have done that, and why." He explained the charges—two counts of first-degree premeditated murder; a special allegation of personal use of a deadly weapon as to both counts; and a special circumstance, which was multiple murder. "Murder PC 187 according to CALJIC 8.10," Bailey explained, "means, 1) A human being was killed. 2) The killing was unlawful. 3) The killing was done with malice aforethought." Then he read the definition of malice aforethought. "Malice is express when there is manifested an intention unlawfully to kill a human being. The mental state constituting malice aforethought does not necessarily require any ill will or hatred of the person killed." He explained that first-degree premeditated murder meant "unlawful killing" and the "willful and deliberate and premeditated express malice." In this case, Bailey argued, the perpetrator had to get the knife. "This is how Rosalie and Amanda were killed," he said, brandishing the weapon, and suggesting that the evidence points to Valentine Underwood as the killer. "He cuts the victims' throats," he said. "Mandi's hands were bound. He raped them. He wants to escape culpability so he commits murders. After the first girl was killed, he had to kill the second. It was obviously willful." Bailey stated that clearly, this was a sexual assault case. Both victims were naked when found. Whole sperm was found in their bodies. The sperm was Valentine Underwood's. "There was the presence of Emotion Lotion at the scene," he said. "It was laying right next to the knife. There was ripped clothing . . . There is an absence of any other motive . . . Who had the motive? Who had the opportunity? Who does the evidence point to? . . . The prints of Valentine Underwood are in blood, in the killing zone, where they shouldn't be . . . Would a reasonable person touch the bodies? Carmichael saw the bodies, he looked, he said, 'Whoa,' left, and called the cops . . . There was a tremendous amount of blood at the crime scene. There were powerful forces—blows—being struck . . . Of the seventy-two stains tested, most are the victims'." But several were the defendant's. Bailey reminded jurors of the exact location of Underwood's blood—on the *Practical English* book near the west wall, where it was mixed with Mandi's; on a towel, mixed with Mandi's; on the sink and tub; at the bottom of the tub; on a razor in a bucket in the tub; on a pink comforter in the bathroom, along with Rosie's blood; on the ripped shower curtain. He moved on to other evidence—Underwood's shoes—showing the soles of the loafers to the jury. The toe print on the knife matched the shoes, he said. "Putting a suspect in actual physical contact with the weapon is very significant," Bailey stated.

In the closing argument for the defense, Hardy suggested that it wasn't so strange for Underwood to have left the crime scene and headed off to a basketball game. "The defendant's life was basketball," Hardy said. "Play ball and we'll get you a degree. Then he joins the Marines and he's running around Twentynine Palms, drinking and chasing women, he's on the basketball team, so it's not a problem." Then he argued that someone else might have been responsible for the double homicide. Trenton Draper, for instance. Perhaps Mandi had impugned Draper's sexuality, Hardy argued. Maybe the reason that Mandi got upset, left Rosie's, and went down the street to visit her Samoan friends was not because men were bothering her but because

Draper had failed to get it up again after she gave him a blow job. Draper was short—about five-eight or nine, Hardy pointed out, suggesting that short men have performance anxiety. When Mandi returned to Rosie's, Draper may have been lying in wait, knife in hand. After killing Mandi, he had to kill Rosie, the eyewitness. He then addressed the DNA evidence, countering the prosecution's contention that Mandi and Rosie had both been raped at or near the time of death by his client. Underwood's sperm had been found in Mandi and Rosie's vaginas and rectums. "How many men can perform four times in one hour?" Hardy said, suggesting that his client had to have had sex with Rosie and Mandi hours before the murders. He paused dramatically after posing the question. The silence begged for a response and one came: the enlarged crime-scene photographs of the mutilated bodies of Mandi and Rosie toppled over, as if offering their own rebuttal. Bailey rose for his final statement. What could he possibly add? everyone wondered. In a movie, this would be the end of the scene, but here in court, it needed embellishment, it must be polished and shined and then handed to the jury like a small treasure of immense weight, a treasure that must be assayed, protected, and returned. He said nothing, tenderly retrieving the crime-scene photographs and replacing them with the photographs he had first shown to the jury. There was Mandi, smiling, cute, with a little bit of baby fat, as she sat for her ninth-grade yearbook picture. There was Rosalie, posing like a model, holding her long hair up on top of her head, smiling, inviting life in for a dance. As the trial came to a close, Mandi and Rosie looked happy again. "Amanda and Rosalie aren't dead," Bailey said with a kind of gratitude, trying not to cry, "they are here. They showed you their nakedness, the sperm inside their bodies, their wounds. They showed you who killed them." In tears and struggling to finish his remarks, after waiting for six years to send the case to a jury, he rested the case of the *People of the State of California* v. *Valentine Underwood*. Debie and Jessielyn wept. The jury filed out, ordered to return the next day to begin deliberations. The defendant's mother left the courtroom, alone.

On December 21, 1997, the Friday before Christmas, Debie and Jessielyn were watching television in Debie's room at the Best Western, waiting for the phone to ring with news that the jury had reached a verdict. News of another verdict interrupted the local broadcast of *All My Children*. Terry Nichols, who had been on trial for conspiring with Timothy McVeigh to blow up the Alfred P. Murrah Federal Building in Oklahoma City, had just been found guilty of involuntary manslaughter. The Mojave lurked in the background of this case, too, and perhaps countless other cases which were playing out across the land, unindicted and unnamed coconspirator in terrible incidents fueled by high-desert nitroglycerine dreams. Debie remarked to Jessielyn that she hoped they would fare as well as the victims of the Oklahoma City bombing and have a verdict before Christmas. The jurors had been weighing the evidence for seven days. They wouldn't want to go home for the holidays and then return in the new year to resume deliberations, would they? It was cold in Victorville, and blustery, with the desert wind blowing right through the motel's doorjamb and the sealant in the window frames, carrying a reminder of death and emptiness and time stopping with no reason to move on. Although the heater was on full blast, it was not able to keep up with the gathering chill. And so Debie and Jessielyn huddled under their winter coats and extra blankets they had piled on the bed, checking their watches and looking to the phone to see if the red light signaling a message had lit up and somehow they had missed it the last time they looked. They were getting nervous. What was hanging them up? they wondered. To them, the case was a slam dunk, and they had gone through at least two packs of cigarettes each since the morning, repeating and convincing themselves of their view. The phone rang. Debie jumped on it. Disappointment—it was Mike, calling from Chicago to find out if there was any news. A little while later, another ring. It was Kat from Escondido, her old friend wondering what was going on. Debie said she'd call her as soon as she heard. About two in the afternoon, the phone rang again. "This has to be it," Jessielyn said. It was. They rushed downstairs to their cars and headed for court.

The jury filed in, not looking to the defendant or the victims. They handed the verdict to Rick the bailiff, who handed it to Bernadette the court clerk. A couple of the cops tensed, eyes trained on Underwood, and then Debie, as Bernadette opened the envelope. "We, the jury in the case of the *People of California* v. *Valentine Underwood,* find the defendant guilty of count one, murder in the first degree of Amanda Lee Scott . . ." Debie pumped a victory gesture with her fist as Bernadette continued. Underwood was found guilty on the other three counts and remanded for sentencing, shaking his head as each verdict was read.

In the hallway, when the jury had filed out, several jurors remained to talk about their deliberations. Theresa Paschal, a respiratory therapist at Loma Linda Veterans Hospital, said that she had tried with all her heart to find something that said the defendant was not at Rosie's at the time of the murders, but could not. The jurors had a hard time believing that Underwood was checking Mandi's pulse. "The wound in her neck was so big," she said. To analyze the event, they put up a time line, she explained, adding that they did not figure out who was killed first. "Obviously," she said, "the second victim was first-degree murder." Heather Hyten, a nurse at Fort Irwin, suggested another scenario. "Maybe he started to kill Mandi," she said, "slashed her throat, and then did Rosie. Mandi was still alive and he had to go back and finish the job." Candy Candelaria had been an occasional patron at the Jolly Roger during the trial. Retired and living in the San Bernardino Mountains, he said that his brother had been in the First Marine Division and was killed at Okinawa, and he was in the Navy during the war in Korea. "It's such a hard thing to do," he said of convicting a fellow Marine. "This man's never going to see daylight again." He explained that the jury had been undecided for a couple of days. A veteran of previous experience as a juror during a rape trial, he employed a strategy he had learned then, suggesting that the jury replay the tape of Underwood's police interview, in which he had lied to police fifty-one times. The foreman, Joe Guiterrez, agreed with Candelaria. "I couldn't believe any of his testimony because of all the lies," he said. An Air Force veteran, Guiterrez was now working on the Marine base at Twentynine Palms as the director of Total Quality Office, a job involved with improving the quality of life aboard the base. "What I had to do first," he said, "was get the jurors to understand how the law applied to the crime. I drew a chart on the board, with two evidence categories—not guilty and guilty. "We could not find a single piece of evidence that would go over on the not-guilty side," he said. "There was no reasonable doubt." Guiterrez added that Hardy's playing of the race card did not work, and in fact was transformed into a joke about Johnnie Cochran, who during the O. J. Simpson trial had coined the phrase "If the glove doesn't fit, you must acquit." Two years after Simpson had been found not guilty, Cochran's line still reverberated in Victorville; the jury deliberating the fate of a poor, unknown black athlete countered with, "His explanation doesn't fit, so we won't acquit." Yet at the very end, Candelaria added, the young black girl on the jury was crying. "She just didn't want to do it to him," he said. The experience of living with this case would take its toll on other jurors in the days to come, just as it had on the key players. Several took to their beds as soon as they got home, gripped by a nasty flu that lingered through the holidays. Guiterrez came down with pneumonia.

Underwood was sentenced the following March. Neither his mother nor his brother attended the hearing. Debie returned from Chicago with Jason and her boyfriend Mike. Jessielyn Gonzalez made the drive from Camp Pendleton. Juanita Brown, her husband Tom, and Rosalie's daughter Shanelle came in from Las Vegas. Lydia Flores joined the families and so did a couple of the jurors. Jessielyn read her statement and so did Kris and Jason; then, finally, after the years of waiting, Debie rose to face her daughter's killer. Again, a few of the cops tightened, hands edging ever so closer to their holsters. "For over six and a half years," she told Underwood, "I have asked everyone, from the DA, the clerks, Rick and other marshals, sheriffs and anyone else, for five minutes, even two minutes alone with you, to no avail. And now after six and a half years, I have my few minutes. Not what I wanted, but I settle for the outcome." To the surprise of spectators, Debie was soft-spoken, her voice even quavering at times. The convicted man conferred with his attorney as she spoke, over her carefully chosen words, as if taunting her, or blocking out the episode. Sometimes, his words were louder than hers and the spectators could barely make out her statement. But she continued.

> *You should have been charged with robbery—because you have robbed two families of their daughters, a daughter of her mother, a mother of her sister, and sisters of their sisters, and grandparents of their granddaughters. I sat in the courtroom for ten weeks and listened without saying a word, absorbing everything—I've had to listen to your attorney Mr. Hardy talk about my daughter Mandi like she was a second-class person—and of course she wasn't here to defend*

herself—like she wasn't allowed to defend herself in death at your hands. Let me tell you both about my daughter Mandi. She was fifteen years old—excited about reaching sweet sixteen. She had a love for life and people. Her memorial service proved that with about two hundred or more people in attendance, of all colors and ages. The whole community cared.

For over six and a half years, your rights under the Constitution and the law have been protected. For me it's been limbo . . . I arrived on the murder scene ten minutes after the sheriffs, paramedics, and fire department, to be told my daughter Mandi was dead. I was not allowed any rights as a parent, but I had to remain outside until three-thirty in the morning, seven and a half hours, answering questions from the sheriff's department so they could do their job. I never got to have that special visit with Mandi. So now after all this time and all these years, I have really learned what patience is all about and it's paid off—'cause now it's your turn to learn what patience is all about—locked up waiting—getting older and older waiting for the day that you die, because justice will and has been served.

Thank you, Mr. Bailey, and your staff for all of your long and hard work. Mr. Fulbright, just for being you. And thanks to a great jury for reaching the right and just verdict. And thanks to this court for enduring this long process—so that our laws of the land could work.

Finally Valentine Underwood rose to make his statement. It would be the last time he would speak in public. He proclaimed his innocence and posed a challenge: "If you actually feel I did it, Your Honor, I don't see why I'm not being put to death."

After years of adjudicating at the proceedings of this trial, Judge Yent delivered a kind of eulogy, as if answering the defendant's question, although, of course, the man about whom and to whom he spoke was very much alive. "Let the punishment fit the crime," he said, citing Dante. "You will never be able to touch a young woman in anger again."

After the verdicts were read, most of the jurors lingered in the hallway outside the courtroom to talk with players in the trial. Debie, in tears, thanked each one and said that for the first time since Mandi had died, she was planning to celebrate Christmas. A couple of jurors expressed concern about the children of Twentynine Palms. What is going on there? they asked. What is going on in our backyard? Debie said that there was a problem in military towns, especially those towns that were "out in the boonies, especially if the boonies means the desert, okay," quickly adding that she had two brothers who were Marines who had served in Vietnam and she was proud of her family's service to the country. But why can't people just leave? the jurors said. Why don't they get out of Twentynine Palms? It's like this, Debie would explain. Have you ever been stuck? I mean really stuck? And the world thinks you're a couple sandwiches shy of a picnic? That's a lot of the people who end up here, okay? Me? Oh no, I left last year. *Hasta la vista*, Two Nine, she would add, adios, see ya, I don't think so, punctuating the reenactment of her departure with a mock military salute.

It was in 1997 that Debie had said farewell to Twentynine Palms, heading far from the desert to Rockford, Illinois, to be near her older daughter, Krisinda, who had moved to Coal Valley with her husband. In a few months, she moved to the suburbs of Chicago to live with her boyfriend Mike, a native of Twentynine Palms who had resettled up north for work and now had an apartment where she and Jason could live. When she left the desert, Debie didn't take much with her. She sold most of her furniture and packed up all of her family scrapbooks and odds and ends and put them in storage. She caught a ride with some friends, bringing Corky and Mandi's urn and a few photos of Mandi, including her favorites—Mandi on the elementary-school swim team; Mandi playing with Corky; Mandi's ninth-grade class picture. A few months later, Jason followed by Greyhound, meeting Debie and Mike in Chicago. Corky, the dog that Mandi had named, and for whom Debie had gone to jail, never made it to Illinois; now fifteen years old, he went into convulsions outside of Casper, Wyoming. Debie took him to a vet and had him put down. He was a desert dog, Debie would tell friends. I don't think he was cut out for the 'burbs. I don't know if I am, but I am doing better up here.

In Chicago, Debie works part-time as a bartender and continues to take in strays. Soon after her arrival, she adopted an abandoned pit bull named Adolf and changed his name to Bubba. He has thrived and calmed some under her care, and has taken a proprietary attitude toward Debie's other adopted animal, a once-neglected iguana named Zoe. During the brutal Midwestern winters, Debie walks the streets and feeds squirrels. During spring, she makes sure to put out feed for the cardinals and blue jays, birds of rich colors not seen in the desert. But the Mojave will not be forgotten. One day not too long ago, the phone rang. It was a long-distance operator, asking if the party wished to accept a collect call from Kevin James. Debie hesitated for a moment, not sure if she wanted to maintain contact with gangbangers in Twentynine Palms, even if this one had been Mandi's boyfriend, then said that she would. "What's up?" she asked. "How'd you get my number?" "You know me," he said, "I got my ways." Then he said hey, he just wanted to let Debie know that his mother had died last week. "Overdose?" Debie asked. "No, AIDS," Kevin said. He missed Mandi. He missed Debie. How was Jason? "By the way," he said, "where you at?" Debie said that Kevin had dialed Chicago. "Man," he said, "you gone for good." That would never happen to him, he said. Most of his homeys were dead or in jail. He figured he didn't have much more time to live: "Be lyin' in chalk pretty soon." Debie told him to watch out for himself. "Didn't Janita have your baby?" she said. "Yeah," Kevin said, "I got a son somewhere. He don't need to see me." "Yeah, he does," Debie said. Kevin said that wasn't going to happen, he was all smoked out and fucked up. "Have you been to the Mandi Tree lately?" Debie asked. "Maybe it's time." "Yeah, maybe," he said, then hung up the phone.

Judge Yent sentenced Valentine Underwood to mandatory life without parole. He was sent to Tehachapi State Prison, and several months later there was an attempt on his life. He was transferred to Corcoran, not an assignment desired by any felon: it is infamous among California penitentiaries as the place where guards pit inmates against each other in fights to the death. He has appealed the guilty verdict and lost. Debie remains embroiled in the legal system. The civil suit which she filed five years ago was thrown out on procedural grounds. She is appealing the dismissal. If the case is reinstated, she is one step closer to a possible trial. If she wins at trial, she hopes that the language on Marine recruitment forms will be forever enforced, or changed. "You are a Marine twenty-four hours a day," it says.

While no one can say if there is a direct connection, ever since the murder of Mandi Scott and Rosalie Ortega, Marines on the base at Twentynine Palms have been discouraged from hanging out in town. To keep them on the base, various amenities have been added: a coffeehouse, a pizza parlor, a video rental store, a couple of new bars, a dry cleaner, a new bowling alley, an in-line skating rink, a theater showing first-run movies.

But the desert echoes with whispers of terrible incidents wrought by Marines who pass time at the Marine Corps Air Ground Combat Center at Twentynine Palms. One story tells of a riot on the base, involving rival gang factions within rival battalions. The riot was so violent that MPs could not keep it under control and the California Highway Patrol was called in to contain it. Other stories tell of trouble in town and all across the eastern California shear zone—knifings, rapes, vehicular manslaughter, child torture, serial murder, a grisly horse massacre; they are told and told again in all the Mojave bars and newspapers, never to be heard elsewhere, swallowed by the very space in which they have been acted.

But, now comes another young girl, to take care of the wounded and the maimed, to minister to those who have been left on the battlefield. In a small desert hospital, a few years after Mandi was killed, a child was born. The baby girl's mother hadn't known Mandi very well, had become part of the magic circle when, in the seventh grade, at a party, she tried to light a cigarette and set her eyebrows on fire. Mandi was the first to help: she put some butter on the burn to calm the pain, and then, to make sure that it never happened again, showed her how to smoke. A few months later, toward the end, Mandi once again rescued her new friend. This time it was from her boyfriend in the school parking lot; during an argument, he had grabbed her and cocked his fist. To Debie's pride and sweet sorrow, the girl called her baby Amanda Lee, in honor of the sweet and fearless schoolmate who had protected her against another black eye. One of the first things she told her daughter was how she got her name.

EPILOGUE

Postscript to a Kill

The jackal that has followed one unlucky family across the generations sleeps again on a pallet of stone. Yet other forces are at work; the concern remains the same. The reference of course is to the Mojave Desert. On October 16, 1999, the Mojave disassembled, cracked open, and sent shock waves of magnitude 7.0 through the ancient lava flows, hollows, and ridges at the dry bed of Lavic Lake on the base at Twentynine Palms. Ordinarily, Lavic Lake is home not to man and woman but to birds and reptiles and insects and wandering feral creatures. But on that night, thousands of Marines played on the sands of the prehistoric lake, carrying out live-fire exercises in the dark. The quake struck like a stealth weapon under the black cover at 0246. There were explosions as the earth heaved off those who scarred and mutilated its flanks in imitation of war. Heavy artillery skittered across the desert gravel like toy guns. Automatic weapons were thrust skyward and then bounced off boulders, firing streams of phosphorescent bullets in all directions. Marines ran for cover but there was none: in the desert, there is no place to hide.

When the sun rose, there came into view a vivid picture. A fault line that had not appeared on maps now splayed across the Mojave for twenty-six miles. On either side of it, the desert floor had moved fifteen feet. Experts walked the strike-slip fissure and ran their fingers across it and whispered in awe and planted listening devices and studied its angles and asked how and why. Their calibrations told them that the fault was young, hadn't existed long enough to alert those who traversed its radius.

A few hours later, the gunfire and bombing runs resumed. But the Mojave had made its point, announced that it had the really heavy artillery, that this little seven-pointer was just a hint of the upheavals that were stored deep in its arsenal, could be deployed when it was time to finally shut the whole thing down. The situation is being closely monitored, as it continues to disturb the military, for there is something it cannot control, a force it cannot vanquish, right in its own backyard, and it is the land of which the world's self-proclaimed most powerful arsenal has taken title. Yet there remains the suspicion, which has lasted longer than the suspicions aroused during previous eruptions, because this one came so close to the very heart of the thing, indeed had erupted perhaps one or two degrees or maybe minutes or seconds even to the west or east of the very heart of the thing, that maybe there needed to be a reassessment somewhere, maybe something had gone terribly wrong. Had the live-fire exercises demanded an answer? Had they themselves triggered a response? Was the earthquake a statement that was eons in the making, a distant echo of another catastrophe that had played out on Mojave's flanks? Of course the suspicion will never be uttered aloud, never appear on a memo and circulate up through the ranks until it sits in a stack of other memos at Marine Corps HQ in Quantico, Virginia. Nevertheless, among those who were blown into the night air by earth's buckling crust, among the enlisted trainees who will at some point in the future defend our oil, our water, our *personal rights*, the message has been received, and classified, to be unearthed at a later date.

NOTES ON THE WRITING OF THIS BOOK

The desert—as it does—took its own sweet time in revealing this story. Although I did not learn of the murders of Mandi Scott and Rosalie Ortega until shortly after the Gulf War in 1991, I heard from the desert long before that. The message was delivered by way of Edgar Allan Poe, thanks to my father, who often read his favorite literary works aloud in his study. One of these was the poem called "Eldorado," Poe's sad tale of the questing knight who forever roamed the sands of time. Although my father died several years ago, I can still hear him reciting the poem, the words drawing me yet again to that sunny and forlorn scape:

Gaily bedight,
A gallant knight,
In sunshine and in shadow,
Had journeyed long,
Singing a song,
In search of Eldorado . . .

Thus began my lifelong involvement with the desert, nourished by my mother, whose love of horses (she was one of the first women in the country to work professionally on the racetrack, as an exercise boy) and early insistence that I learn to ride and explore the world showed me that the world looks better from atop a horse, even if the world at that time was northeastern Ohio instead of the one that is marked by cactus and sage, where I have since spent much of my time.

When I cannot get to the grail itself, I find myself immersed in works which echo, broaden, or deepen Poe's rhyme, which tell in a particular way of what Shakespeare called "the brief walk through nature to eternity." So, in addition to my parents, I must give credit to the foundation for this book, my many sources of inspiration—fiction, nonfiction, essays, movies, and music alike—which include, not in any particular order: *The Land of Little Rain* by Mary Austen; anything by Wallace Stegner but most especially *Angle of Repose* and his essay "The American West As Living Space"; *Unassigned Territory* by Kem Nunn; anything by Barry Lopez; anything by John Steinbeck, especially *Of Mice and Men; America* by Jean Baudrillard; the poem "Coming into Animal Presence" by Denise Levertov; the works of Ursula K. Le Guin; the film *Badlands* by Terence Malick; *The Treasure of the Sierra Madre* by B. Traven, and of course, the film adaptation by John Huston; and, among many other sources too numerous to mention, a gift of music which is surely a kind of radio transmission from the land of little rain itself—"Journey in Satchidananda" by Alice Coltrane.

As the desert bared its revelation came the problem of how to shape it. These are the works which both inspired and served as literal keys to the construction of this narrative: *Shot in the Heart* by Mikal Gilmore; *Common Ground* by J. Anthony Lukas; *All God's Children* by Fox Butterfield; *The Living and the Dead* by Paul Hendrickson; *The Executioner's Song* by Norman Mailer; and, of course, *In Cold Blood* by Truman Capote.

And then there are the books which served as primary

research. For "official" history, I consulted the following: *California: An Illustrated History* by T. H. Watkins; *Southern California: An Island on the Land* by Carey McWilliams; Kevin Starr's three-part *California Dream* series; *1500 California Place Names* by William Bright; and *A Guide to Rock Art Sites: Southern California and Southern Nevada* by David S. Whitley. For natural history and information about the terrain, I consulted *Simon & Schuster's Guide to Cacti and Succulents* edited by Stanley Schuler; *The California Deserts* by Edmund C. Jaeger; *A Peculiar Piece of Desert: The Story of California's Morongo Basin* by Lulu Rasmussen O'Neal; *Desert Lore of Southern California* by Choral Pepper; *Deserts* by James A. MacMahon, an Audubon Society Nature Guide; *California's Changing Landscapes (Diversity and Conservation of California Vegetation)* by Michael Barbour, Bruce Pavlik, Frank Drysdale, and Susan Lindstrom; various reports from the United States Geological Survey; and a UCLA anthropology-department Marine-commissioned study of Native American art on the base at Twentynine Palms. For accounts of the early settlers of Twentynine Palms, I referred to the two-volume series *In the Shadow of the Palms* by Art Kidwell, and for Native American mythology and history, to *California Indian Nights; The Earth Is Our Mother: A Guide to the Indians of California: Their Locales and Historic Sites;* and brochures number one through three from the Malki Museum on the Morongo Indian Reservation about the Cahuilla, the Serrano, and the Chemehuevi Indians of Southern California. For Marine history and lore, I consulted *Semper Fidelis: The History of the United States Marine Corps* by Allan R. Millett; *Marine! The Life of Chesty Puller* by Burke Davis; *330 Reasons to Love the Corps* written and compiled by Chris Lawson; and *U.S. Marines at Twentynine Palms, California* by Colonel Verle E. Ludwig, U.S. Marine Corps, Retired. For background about the children of the region, I read the *Third Annual*

Report from the Children's Network of San Bernardino County: Children at Risk and Children's Agenda, '91–'92, and for background of a different sort, I found *Bury Me Standing: The Gypsies and Their Journey* by Isabel Fonseca quite helpful.

Since 1990, I have subscribed off and on to various desert newspapers, including *The San Bernardino Sun, The Hi-Desert Star, The Desert Trail.* I like receiving these newspapers in Los Angeles—it helps to sustain my connection with the desert on the days when city life nearly obliterates memory of space. After I began investigating the murders, I subscribed to *Navy Times—The Marine Corps Edition, Marine Corps Gazette,* and *Leatherneck—Magazine of the Marines.* The news and information published in these newspapers and magazines have been essential to my understanding of this story. In fact, not having been in Twentynine Palms at the time of the murders and missing the brief flurry of local news coverage, I would never have heard anything other than bar tales about the incident had it not been for an article by Gregg Patton in *The San Bernardino Sun* in which I first learned the names of the victims and that the mother of one of them, Debie McMaster, tended bar somewhere in the Mojave.

Other sources of published material include the San Bernardino County coroner's report; San Bernardino County Sheriff's Department police reports and interviews; the transcript of the preliminary hearing at Joshua Tree, California; transcripts related to motions argued at court in Barstow, California; transcripts of the murder trial at Victorville, California; and the official, although partially blacked-out Naval Criminal Investigative Service (NCIS) investigation of Valentine Underwood, as obtained via the Freedom of Information Act (FOIA).

Of course, the writing of this book depended most of all on interviews with the players in this tragedy. First, I owe lifelong gratitude to Debie McMaster, who let me into her life, and world, so that I could tell the story of the legacy of violence which has plagued her family, and how it caught up with Mandi, who always came to the aid of her beleaguered friends. I am grateful also to Debie's son Jason Scott and surviving daughter Krisinda Fuselier, as well as to her mother Rose Powell, father Clarence McMaster, brother Darrell McMaster, cousin and family historian James Prunetti, and longtime companion Mike Ramirez. I am also indebted to Jessielyn Gonzalez for helping me get to know her sister Rosalie, and to Rosalie's mother Juanita Brown, her daughter Shanelle, and Juanita's husband Tom. Among the key players, I am most thankful to—and continually amazed by—Mandi and Rosie themselves, for living such incredible and heroic lives amid a landscape of extreme emotional and social desolation.

I must also thank Tammy Watson for revisiting a horrible incident which she had tried hard to forget, and for putting me in touch with her father George Watson, a former sergeant major in the Marine Corps, who was gracious in our two brief conversations about his daughter's ordeal.

I am also grateful to Zimena Underwood and her son Tracy, although my conversations with them were not extensive.

To my regret, despite several attempts, I was not able to discuss certain episodes in Debie McMaster's second marriage with Max Scott. Nor was I able to reach Debie's first husband, Carl Fuselier.

I would also like to acknowledge the contributions of the girls in the Lunch Box Gang, primarily Lydia Flores, Sandy Flores, and Tina Herrera; former Marine Scott Stewart and his wife Laurie, both proprietors of the now-defunct Twentynine Palms coffee house the Casa de Java. Vickie Waite, publisher of a local monthly called *The Sun Runner* for background on Twentynine Palms; Timothy Carmichael for his account of being on the front in the Gulf War; the cops who gave me their thoughts on the incident, especially Detective Norm Parent and Sergeant Tom Neely; Captain John Manley of the USMC for information about the Corps; former Marine Jack Shea for background on life in the Corps; former Marine R. J. Thrasher for his account of Marine training and life, as well as for aid in Twentynine Palms as I moved through the nooks and crannies; Rick Hertberg of the NCIS for facts about the Marine investigation of Valentine Underwood; Valentine Underwood's second lawyer Garrett Zelen for his background on the case; San Bernardino County district attorney Gary Bailey, victim witness advocate Jesse Fulbright, and the Victorville court staff, including the bailiffs and court reporters and clerks, who have been most gracious over the years; the jurors who spoke with me after the verdict was handed down; and the countless others who let me into their lives so I could write this book, some of whom have requested anonymity (pseudonyms are indicated by the phrase "known as" or "went by the name of"), including the gangbangers who called collect from jail to spin tales of their hopeless lives; the drunks in desert bars who are ashamed of themselves and their plight and their bloodshot eyes, so they hide from the sun, and the rest of the world; enlisted Marines—good and bad—for talking about a job, a life, that is scorned and avoided by opinion makers, regardless of the occasional Hollywood tip of the hat, and especially the children of the desert, who, for better or for worse, know that they are headed nowhere.

In writing this book, help has also come from other sources in other ways. Certain people have been key along the way, including my dear cousin Jon Stillman and his generous wife Denise, my friends Steve and Jeri de Souza, Ruth Charny, Amy Handelsman, Nancy Hardin, Tom Teicholz, Hank Wise, Lisa Kirk, Willard Morgan, Larry Gogolick, Sherri Foxman, Mark Rogers, Carol Rogers, Eric Feig (also an adviser and my attorney), Leslie Caveny, Louie Liberti, Peggy Garrity, and John Angello. Many thanks also to the William Morrow staff—crack Associate Editor Sarah Durand, and diligent publicist Kristen Green and her assistant Andrea Pfaundler. To all whose names I've failed to mention, please know that no slight is intended; your encouragement and support over the years have been beyond measure.

In addition, I am grateful to filmmaker Nina Menkes, whose own research into a murder in Twentynine Palms unearthed the Gulf War marching cadence which I have found very useful to my work. I would also like to thank Mark Horowitz, Michael Caruso, and Bob Roe, who published my article "Murder at Twentynine Palms" when they were editors at *Los Angeles* magazine in April 1996.

Finally, I must acknowledge Sonja Bolle for her invaluable editorial guidance; early advocates Rachel Klayman and Hamilton Cain; my agent Kathy Anderson for helping me take my work down this path; film agent Bill Contardi for giving my manuscript safe passage through Hollywood, and my William Morrow editor Trish Grader for waiting patiently as I missed several deadlines while journeying through the Mojave, and for her passion for my book when I finally wandered in for a rest.

"Shadow," said he
"Where can it be—
This land called Eldorado? . . ."